Information and Instructio

This shop manual contains several sections each covering a specific group of wheel type tractors. The Tab Index on the preceding page can be used to locate the section pertaining to each group of tractors. Each section contains the necessary specifications and the brief but terse procedural data needed by a mechanic when repairing a tractor on which he has had no previous actual experience.

Within each section, the material is arranged in a systematic order beginning with an index which is followed immediately by a Table of Condensed Service Specifications. These specifications include dimensions, fits, clearances and timing instructions. Next in order of arrangement is the procedures paragraphs.

In the procedures paragraphs, the order of presentation starts with the front axle system and steering and proceeding toward the rear axle. The last paragraphs are devoted to the power take-off and power lift systems. Interspersed wh
specifications pertain

HOW TO USE THE INDEX

Suppose you want to know the procedure for R&R (remove and reinstall) of the engine camshaft. Your first step is to look in the index under the main heading of ENGINE until you find the entry "Camshaft." Now read to the right where under the column covering the tractor you are repairing, you will find a number which indicates the beginning paragraph pertaining to the camshaft. To locate this wanted paragraph in the manual, turn the pages until the running index appearing on the top outside corner of each page contains the number you are seeking. In this paragraph you will find the information concerning the removal of the camshaft.

More information available at haynes.com
Phone: 805-498-6703

Haynes UK
Sparkford Nr Yeovil
Somerset BA22 7JJ England

Haynes North America, Inc
859 Lawrence Drive
Newbury Park
California 91320 USA

ISBN-10: 0-87288-106-7
ISBN-13: 978-0-87288-106-8

© **Haynes North America, Inc. 1964**
With permission from J.H. Haynes & Co. Ltd.

Clymer is a registered trademark of Haynes North America, Inc.

Cover art by Sean Keenan

Disclaimer

There are risks associated with automotive repairs. The ability to make repairs depends on the individual's skill, experience and proper tools. Individuals should act with due care and acknowledge and assume the risk of performing automotive repairs.

The purpose of this manual is to provide comprehensive, useful and accessible automotive repair information, to help you get the best value from your vehicle. However, this manual is not a substitute for a professional certified technician or mechanic.

This repair manual is produced by a third party and is not associated with an individual vehicle manufacturer. If there is any doubt or discrepancy between this manual and the owner's manual or the factory service manual, please refer to the factory service manual or seek assistance from a professional certified technician or mechanic.

Even though we have prepared this manual with extreme care and every attempt is made to ensure that the information in this manual is correct, neither the publisher nor the author can accept responsibility for loss, damage or injury caused by any errors in, or omissions from, the information given.

SHOP MANUAL
INTERNATIONAL HARVESTER

SERIES 460-560-606-660-2606

Engine serial number is stamped on left side of engine crankcase. Engine serial number will be preceeded by engine model number. Suffix letters to engine serial number are as follows:

C. LP-Gas Burning Engine

U. High Altitude Engine

Tractor serial number is stamped on name plate attached to right side of clutch housing. Suffix letters to tractor serial numbers indicate following attachments:

C. LP-Gas (Standard Altitude)
F. Cotton Picker Mounting (High Drum)
G. Cotton Picker Tractor Attachment (High Drum)
H. Rear Frame Cover and Gear Shifter
J. Rockford Clutch
P. Independent PTO Drive (540 rpm)
S. "Torque Amplifier" With Provision for 540 R.P.M. Independent PTO

T. Cotton Picker Mounting (Low Drum)
U. High-Altitude
W. Forward and Reverse Drive
X. High Speed Low and Reverse
Y. Hydraulic Power Supply (12 gpm pump)
Z. Hydraulic Power Supply (17 gpm pump)
CC. 3rd Speed Heavy Tillage Gear
FF. Hydraulic Power Supply (4.5 gpm pump)
GG. Hydraulic Power Supply (7.0 gpm pump)

INDEX (By Starting Paragraph)

CONDENSED SERVICE DATA

GENERAL	Non-Diesel 460	Diesel 460	Non-Diesel 560, 660	Diesel 560, 660	Non-Diesel 606, 2606	Diesel 606, 2606
Engine Make	Own	Own	Own	Own	Own	Own
Engine Model	C-221	D-236	C-263	D-282	C-221	D-236
Number of Cylinders	6	6	6	6	6	6
Bore-Inches	$3\frac{9}{16}$	$3\frac{11}{16}$	$3\frac{9}{16}$	$3\frac{11}{16}$	$3\frac{9}{16}$	$3\frac{11}{16}$
Stroke-Inches	$3\frac{11}{16}$	$3\frac{11}{16}$	4.39	4.39	$3\frac{11}{16}$	$3\frac{11}{16}$
Displacement—Cubic Inches	221	236	263	282	221	236
Main Bearings, Number of	5	5	5	5	5	5
Cylinder Sleeves	Dry	Dry	Dry	Dry	Dry	Dry
Forward Speeds, No T.A.	5	5	5	5	5	5
Forward Speeds, With T.A.	10	10	10	10	10	10
Generator and Starter Make	Delco-Remy					

TUNE-UP						
Compression Pressure Except LPG	155*	350*	155*	375*	155*	350*
LPG	205*	....	205*	....	205*	....
Firing Order	1-5-3-6-2-4					
Valve Tappet Gap (Hot)	0.027	0.027	0.027	0.027	0.027	0.027
Inlet Valve Seat Angle (Degrees)	30	45	30	45	30	45
Exhaust Valve Seat Angle (Degrees)	30	45	30	45	30	45
Ignition Distributor Make	IH	....	IH	....	IH	....
Ignition Distributor Symbol	See Par. 181	....	See Par. 181	....	See Par. 181	....
Breaker Contact Gap	0.020	....	0.020	....	0.020	....
Distributor Timing	See Par. 180	....	See Par. 180	....	See Par. 180	....
Timing Mark Location	Crankshaft Pulley					
Spark Plug Electrode Gap (Gasoline)	0.023	....	0.023	....	0.023	....
LP-Gas	0.015	....	0.015	....	0.015	....
Carburetor Make, Gasoline	IH	....	IH	....	IH	....
Carburetor Model, Gasoline	$1\frac{1}{4}$	....	$1\frac{3}{8}$	....	$1\frac{1}{4}$	....
Carburetor Make, LP-Gas	Ensign	....	Ensign	....	Ensign	....
Carburetor Model, LP-Gas	XG	....	XG	....	XG	....
Battery Terminal Grounded	Negative	Negative	Negative	Negative	Negative	Negative
Engine Low Idle RPM	425	650	425	650	425	650
Engine High Idle RPM	1975	1965	1975(1)	1965(1)	2200	2180
Engine Governed RPM	1800	1800	1800(1)	1800(1)	2000	2000

(1) Applies to series 560. Series 660 non-diesel high idle rpm, 2640; diesel high idle rpm, 2615. Governed rpm for both engines, 2400 rpm.

*Approximate psi, at sea level, at cranking speed.

SIZES—CLEARANCES—CAPACITIES
(Clearances in Thousandths)

	Non-Diesel 460	Diesel 460	Non-Diesel 560, 660	Diesel 560, 660	Non-Diesel 606, 2606	Diesel 606, 2606
Crankshaft Main Journal Diameter	2.7485	2.7485	2.7485	2.7485	2.7485	2.7485
Crankpin Diameter	2.3735	2.3735	2.3735	2.3735	2.3735	2.3735
Camshaft Journal Diameter, No. 1 (Front)	2.1095	2.1095	2.1095	2.1095	2.1095	2.1095
Camshaft Journal Diameter, No. 2	2.0895	2.0895	2.0895	2.0895	2.0895	2.0895
Camshaft Journal Diameter, No. 3	2.0695	2.0695	2.0695	2.0695	2.0695	2.0695
Camshaft Journal Diameter, No. 4	1.5000	1.5000	1.5000	1.5000	1.5000	1.5000
Piston Pin Diameter, Diesel (early)	....	0.87485	....	0.87485	....	....
Diesel (late)		1.1248		1.1248		1.1248
Non-Diesel	0.87485	....	0.87485	....	0.87485	....
Valve Stem Diameter	0.372	0.372	0.372	0.372	0.372	0.372
Main Bearing Diametral Clearance	1.2-4.2	1.2-4.2	1.2-4.2	1.2-4.2	1.2-4.2	1.2-4.2
Rod Bearing Diametral Clearance	.9-3.4	.9-3.4	.9-3.4	.9-3.4	.9-3.4	.9-3.4
Piston Skirt Diametral Clearance	2-3	4-5.6	2-3	4-5.6	2-3	4-5.6
Crankshaft End Play	5-13	5-13	5-13	5-13	5-13	5-13
Camshaft Bearings Diametral Clearance	.5-5	.5-5	.5-5	.5-5	.5-5	.5-5
Camshaft End Play	2-10	2-10	2-10	2-10	2-10	2-10
Cooling System Capacity, Quarts	$20\frac{1}{2}$	23	$20\frac{1}{2}$	23	$20\frac{1}{2}$	23
Crankcase Oil, Quarts	9	9	9	9	9	9
PTO Rear Unit, Quarts:						
Planetary Type	2	2	2	2	2	2
Clutch Type					$\frac{1}{2}$	$\frac{1}{2}$
Transmission and Differential, Gals.	10	10	16	16	13	13

TIGHTENING TORQUES—FOOT POUNDS

	Non-Diesel 460	Diesel 460	Non-Diesel 560, 660	Diesel 560, 660	Non-Diesel 606, 2606	Diesel 606, 2606
Camshaft Nut	115	115	115	115	115	115
Rod Bearing Screws	50	50	50	50	50	50
Cylinder Head Screws	90	115	90	115	90	115
Flywheel Screws	60	60	60	60	60	60
Injection Nozzle Hold Down Screws	....	20-25	....	20-25	....	20-25
Injection Nozzle Fitting		65		65		65
Main Bearing Screws	80	80	80	80	80	80
Manifold Screws	25	25	25	25	25	25

3

FRONT SYSTEM-TRICYCLE TYPE

DUAL FRONT WHEELS

Series 460-560

1. Dual wheel tricycle front wheels are available on Farmall 460 and 560 tractors. The lower bolster (45—Fig. IH1000) is bolted directly to the upper bolster pivot shaft (42). The wheel axles are riveted to the lower bolster and are available separately. Wheel bearings should be adjusted to a slight pre-load by turning the bearing adjusting nut (16 — Fig. IH-1001).

SINGLE FRONT WHEEL

Series 460-560

2. Farmall 460 and 560 tractors are available with a single wheel tricycle front wheel. The wheel fork (2—Fig.

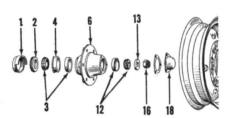

Fig. IH1001—Exploded view of the front wheel hub and bearings.

1. Dust shield	12. Outer bearing
2. Oil seal	13. Washer
3. Inner bearing	16. Nut
4. Oil seal retainer	18. Cap
6. Hub	

IH1002) is bolted directly to the upper bolster pivot shaft (42—Fig. IH1000).

Two types of wheel and axle assemblies (Fig. IH1002) are available.

One type of wheel and axle assembly has a conventional wheel (13), solid hub (12) and axle shaft (6) with tapered roller bearings (8 and 10). The wheel bearings can be adjusted to a slight pre-load by turning the bearing adjusting nut (17). Adjustment is locked by nut (17N).

The other type wheel and axle assembly uses a wheel (21) which consists of two halves (male and female). The wheel halves of early models are equipped with renewable bushings (20) and rotate on shaft (29). Side play of wheel is adjusted by nuts (23) and locked by lock washers (24). wheel halves of late models are equipped with tapered roller bearings.

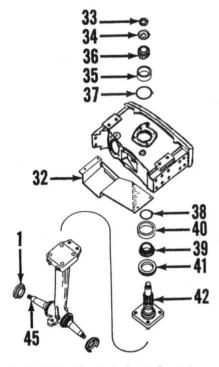

Fig. IH1000—The dual wheel tricycle lower bolster (45) is attached to the upper bolster pivot shaft (42).

1. Dust shield	37 & 38. "O" rings
32. Crankshaft pulley shield	39. Bearing cone
	40. Bearing cup
33. Nut	41. Oil seal
34. Lock washer	42. Upper bolster pivot shaft
35. Bearing cup	
36. Bearing cone	45. Lower bolster

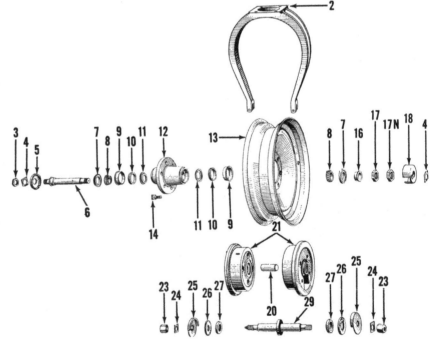

Fig. IH1002—Exploded view shows both types of fork mounted single front wheels. Axle (6) is carried in taper roller bearings; whereas, axle (29) is carried in two bushings (20).

2. Wheel fork	9. Oil seal retainer	17. Bearing adjusting nut	23. Nut
3. Nut	10. Bearing cup	17N. Jam nut	24. Lock washer
4. Lock washer	11. Grease retainer	18. Dust shield	25. Shield
5. Dust shield	12. Hub	20. Bushing (2 used)	26. Felt washer
6. Axle	13. Wheel	21. Wheel halves	27. Oil seal
7. Oil seal	16. Spacer		29. Axle
8. Bearing cone			

FRONT SYSTEM-AXLE TYPE

AXLE MAIN MEMBER

Farmall 460-560 Adjustable Axle

3. On adjustable axle models shown in Figs. IH1003 and IH1004, the main member pivots on pin (21) which is pinned in the adapter or lower bolster (19). Two pivot pin bushings (12), which are pressed into the main member, should be reamed after installation to provide a free fit for the pin.

Series 460-606-2606 International

4. On 460-606-2606 International tractors equipped with either an adjustable or non-adjustable axle (Figs. IH1005 or IH1006), the radius (stay) rod is welded to the axle main member and is not available separately. The front axle pivots on pin (14) which is carried in bushing (21). Pin (14) is bolted to the front axle pivot pin bracket (20). Ream bushing (21), if necessary, to provide a free fit for pin (14).

On 606 and 2606 International tractors equipped with either an adjustable or non-adjustable axle (Figs. IH1006A or IH1006B), the radius (stay) rod is welded to the axle main member and is not available separtely. The front axle pivots on pin (5) which is carried in bushing (6). Pin is retained in front support (10) and bracket (7) by lock plate (25). Ream bushing (6), if necessary, to an inside diameter of 1.751-1.755 to provide an operating clearance of 0.001-0.007 for the 1.748-1.750 diameter pivot pin.

Series 560-660 International

5. The front axle (21—Fig. IH1007) pivots on pin (15) which is carried in bushings (22). New bushings may need to be reamed after installation to provide a free fit for pivot pin (15).

RADIUS (STAY) ROD AND PIVOT

Farmall 460-560 Adjustable Axle

6. The radius rod pivot ball (16—Fig. IH1003 or IH1004) is available for service as an individual part. Clearance between pivot ball and socket can be adjusted by the addition or subtraction of shims (11). The radius rod (14) can be detached from the axle (33).

Series 460-606-2606 International

7. The radius rod pivot ball (22—Fig. IH1005 or IH1006) or (26—Fig.

IH1006A or 1006B) is available for service as an individual part. Diametral clearance between pivot ball and socket should be 0.005-0.020 and can be adjusted by the addition or subtraction

of shims (12 or 3). The ball socket cap retaining cap screws should be tightened to 120-135 ft.-lbs. of torque. The radius rod is welded to the axle and is not detachable.

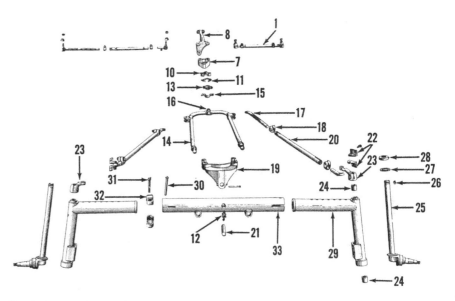

Fig. IH1003—Adjustable axle and components used on 460 and 560 Farmall "Hi-Clear" models. Refer to Fig. IH1004 for legend.

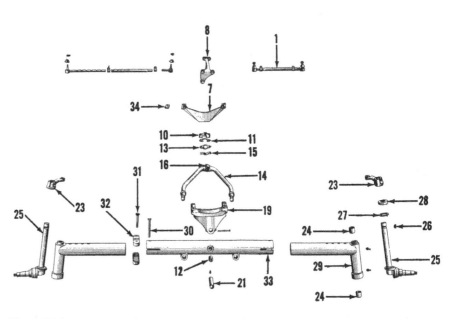

Fig. IH1004—Exploded view of the special adjustable front axle available for Farmall 460 and 560 tractors.

1. Tie rod assembly	12. Bushings	21. Axle pivot pin	28. Thrust bearing
7. Ball socket support	13. Ball socket cap	23. Steering knuckle arm	29. Axle extension
8. Steering gear arm	14. Stay rod	24. Bushing	30. Clamp pin
10. Ball socket	15. Lock plate	25. Steering knuckle	31. Clamp bolt
11. Shim	16. Stay rod ball	26. Woodruff key	32. Axle clamp
	18. Clamp	27. Felt washer	33. Axle center member
	19. Lower bolster		

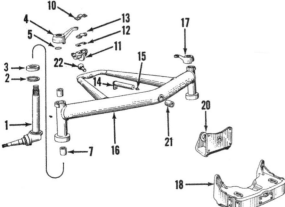

Fig. IH1005—International 460 adjustable front axle, pivot bracket and associated parts.

1. Knuckle	7. Bushings	12. Shims	17. Left steering arm
2. Felt washer	8. Axle extensions	13. Ball cap	18. Bolster
3. Thrust bearing	9. Axle clamp	14. Pivot pin	20. Pivot bracket
4. Right steering arm	11. Ball bracket	16. Front axle	21. Bushing
5. Snap ring			22. Pivot ball

justable for wear. On 560 and 660 International tractors, the tie rod is equipped with automotive type ends; however, the drag link uses an adjustable socket and ball on the forward end and a block and yoke type joint at the rear end. Although the drag link differs on power steering models, both ends are the same as those used on the manual steering drag link shown in Fig. IH1020.

Adjust the toe-in of the front wheels to ⅛-¼-inch on series 460 and 560 Farmall adjustable axle type tractors, ¼-⅜-inch on series 560 and 660 International non-adjustable axle type tractors and $\frac{3}{16}$-⅜-inch on 460, 606 and 2606 International adjustable and non-adjustable axle tractors. Adjustment is made by varying the length of the tie rods or drag links. On some 460 International tractors, alignment marks are provided on the steering arms and axle extensions as well as on the Pitman arms and gear housing to facilitate making the toe-

Fig. IH1006 — Exploded view of the non-adjustable axle used on some 460 International tractors. Refer to Fig. IH1005 for legend.

Series 560-660 International

8. The radius rod pivot ball (5—Fig. IH1007) is available for service as an individual part. Clearance between pivot ball and socket can be adjusted by the addition or subtraction of shims (3). The radius rod (6) is detachable from the axle main member (21).

TIE RODS AND DRAG LINKS
All Models So Equipped

9. Procedure for removing the tie rod or rods, tie rod tubes, tie rod ends and/or drag links is self-evident after an examination of the unit. The tie rod and drag link ends, on all except International 560 and 660 tractors, are of the automotive type and are not ad-

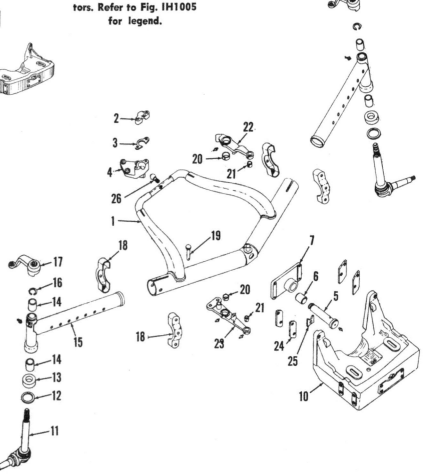

1. Front axle	16. Snap ring
2. Ball cap	17. Right steering arm
3. Shims	18. Axle clamp
4. Ball bracket	19. Cylinder bracket pin
5. Pivot pin	20. Bushing (large)
6. Bushing	21. Bushing (small)
7. Pivot bracket	22. Cylinder support
10. Bolster	arm (upper)
11. Knuckle	23. Cylinder support
12. Felt washer	arm (lower)
13. Thrust bearing	24. Axle stop
14. Bushing	25. Lock plate
15. Axle extension	26. Pivot ball

Fig. IH1006A—Exploded view of the adjustable front axle used on series 606 and 2606 International tractors. Refer also to Fig. IH1006B.

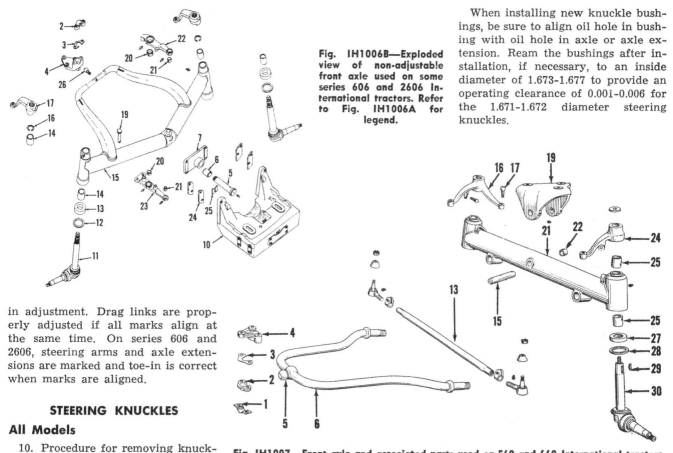

Fig. IH1006B—Exploded view of non-adjustable front axle used on some series 606 and 2606 International tractors. Refer to Fig. IH1006A for legend.

When installing new knuckle bushings, be sure to align oil hole in bushing with oil hole in axle or axle extension. Ream the bushings after installation, if necessary, to an inside diameter of 1.673-1.677 to provide an operating clearance of 0.001-0.006 for the 1.671-1.672 diameter steering knuckles.

in adjustment. Drag links are properly adjusted if all marks align at the same time. On series 606 and 2606, steering arms and axle extensions are marked and toe-in is correct when marks are aligned.

STEERING KNUCKLES

All Models

10. Procedure for removing knuckles from axle or axle extensions is evident after an examination of the unit and reference to Fig. IH1003, 1004, 1005, 1006, 1006A, 1006B or 1007.

Fig. IH1007—Front axle and associated parts used on 560 and 660 International tractors.

1. Lock plate	6. Stay rod	19. Pivot bracket	25. Bushings
2. Ball cap	13. Tie rod	21. Axle	27. Thrust bearing
3. Shim	15. Axle pivot pin	22. Bushings	28. Felt washer
4. Ball socket	16. Left steering arm	24. Right steering	29. Woodruff key
5. Stay rod ball	17. Ball stud	arm	30. Knuckle

MANUAL STEERING SYSTEM (FARMALL)

ADJUSTMENTS. The manual steering system used on series 460 and 560 Farmall tractors is provided with the following two adjustments.

12. SECTOR (UPPER BOLSTER PIVOT) SHAFT END PLAY. The sector shaft end play can be adjusted as follows: Remove the radiator as outlined in paragraph 175. Support the tractor in such a way as to remove weight from the front wheels. Remove the grille and the cap screws which retain the radiator bracket to steering gear housing. NOTE: Four of the radiator bracket retaining cap screws can be removed from above, the other four from below. Lift the radiator bracket from tractor.

The adjusting nut (33—Fig. IH1011) should be tightened sufficiently to provide the bearings with a slight preload.

Fig. IH1011—The upper bolster pivot shaft end play on Farmall tractors is adjusted by turning nut (33).

13. WORM SHAFT END PLAY. The worm shaft end play can be adjusted as follows: Support the tractor in a suitable manner and remove the engine left side rail. Unbolt the steering shaft bracket from the clutch housing and slide the front universal joint from the worm shaft. Unbolt the end cover

(15—Fig. IH1015) from the steering gear housing. Remove the end cover making certain that seals (16 and 19) are not damaged. Unstake nut (13) and tighten until all bearing end play is removed without causing any binding tendency. NOTE: In most cases it will be necessary to renew nut; also, the threads on the worm shaft should be inspected for damage. In some cases excessive end play may be due to the set screw (2) being loose in the worm shaft nut. Refer to paragraph 16 to remove rack and tighten set screw.

STEERING GEAR

14. **REMOVE AND REINSTALL.** Removal of the complete steering gear and radiator assembly is necessary for jobs requiring removal of the crankshaft pulley, timing gear cover and others.

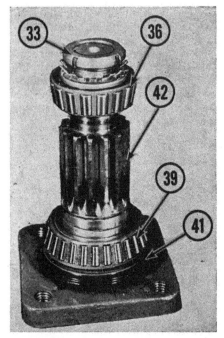

Fig. IH1012—The upper bolster pivot shaft (42) and associated parts. Care must be taken to prevent damage to the seal (41).

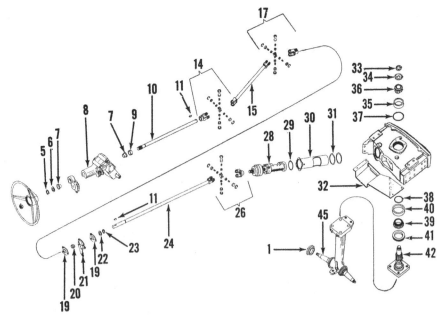

Fig. IH1013—Exploded view of Farmall steering system. The worm unit (28) is shown exploded in Fig. IH1015.

1. Dust shield	15. Steering shaft	26. Universal joint	37. "O" ring
5. Snap ring	(center)	28. Worm unit	38. "O" ring
6. Washer	17. Universal joint	29. "O" ring	39. Bearing cone
7. Bearings	19. Bearing retainers	30. Sleeve	40. Bearing cup
8. Shaft housing	20. Bearing	31. "O" rings	41. Oil seal
9. Shaft collar	21. Bearing support	32. Crankshaft	42. Upper bolster
10. Steering shaft	bracket	pulley shield	pivot shaft
(rear)	22. Thrust washer	33. Adjusting nut	45. Dual wheel
11. Woodruff keys	23. Snap ring	34. Lock washer	tricycle lower
14. Universal joint	24. Steering shaft	35. Bearing cup	bolster
	(front)	36. Bearing cone	

To remove the steering gear and radiator as an assembly, first remove the four hood side sheets (skirts) and unbolt the front of the hood from the radiator bracket. Drain coolant and disconnect the radiator hoses. Support the tractor under the clutch housing and sling the steering gear and radiator assembly in a hoist. Remove the engine left side rail and the cap screws which retain the right side rail to the steering gear housing (front bolster). Move the radiator and steering assembly forward away from the tractor.

When reinstalling, reverse the removal procedure. Make certain that steering shaft front universal joint is guided on worm shaft as the radiator and steering assembly is rejoined to the tractor.

OVERHAUL. Although most repair jobs associated with the steering gear will require the removal and overhaul of both the sector (upper bolster pivot) shaft and the steering worm, there are some instances where the failed or worn part is so located that repair can be accomplished by removing only one. In effecting such localized repairs, time will be saved by observing the following two paragraphs as a general guide.

15. SECTOR (UPPER BOLSTER PIVOT) SHAFT. The sector shaft (42 —Fig. IH1013) can be removed for

service as follows: Remove the radiator as outlined in paragraph 175. Support tractor in such a way as to remove weight from the front wheels. Remove grille and the cap screws which retain radiator bracket to steering gear housing. NOTE: Four of the radiator bracket retaining cap screws can be removed from below the other four can be removed from above. Lift radiator bracket from tractor. On adjustable axle equipped tractors, unbolt the radius rod pivot bracket from clutch housing. On all models, bend the locking tab from the adjusting nut (33—Fig. IH1011); then, remove the nut. Lift the front of the tractor up enough to clear the top of the sector shaft and roll wheels and sector shaft away from tractor.

The lower bearing cone (39—Fig. IH1012) and seal (41) can be pulled from sector shaft (42) and cups (35 and 40—Fig. IH1013) can be pulled from gear housing if renewal is required.

When reassembling, renew all "O" rings and seals and mesh the line marked tooth on rack with valley between the two dotted teeth of sector gear. Refer to Fig. IH1014 for view of timing marks. The bearing adjusting

Fig. IH1014 — When installing the upper bolster pivot shaft, align the timing marks as shown.

nut should be tightened to provide bearings with a slight pre-load.

16. STEERING WORM UNIT. To remove the steering worm, first remove all front end weights attached to the steering gear housing; then, remove the sector shaft as outlined in paragraph 15. Remove the engine left side rail, unbolt the steering shaft bracket from the clutch housing and pull the steering shaft front universal joint from the worm shaft. Unbolt the worm shaft end cover from the gear housing and loosen the four cap screws which attach the housing to the right side rail enough to permit withdrawal of the worm, nut, rack and end cover assembly.

Disassembly is self-evident after an examination of the unit and reference to Fig. IH1015. When removing worm and nut (7), unstake and unscrew set screw (2) and withdraw nut and worm from rack. Seal (16) should be installed with lip inward. Needle thrust bearings (9) and races (8 and 11) should be renewed if they are in any way questionable. Nut (13) should be tightened to just remove all bearing end play, then stake nut to maintain this adjustment. The worm nut is re-

Fig. IH1015 — Exploded view of Farmall manual steering worm unit.

1. Rack	6. "O" ring
2. Set screw	7. Worm and nut
3. Rack plate	8. Bearing races
4. Special screw	9. Thrust bearings
5. Adaptor	10. Washers

11. Bearing race	16. Oil seal
12. Lock washer	17. Bearing
13. Lock nut	18. Washer
15. Cover	19. Oil seal

tained in rack (1) by the cone point set screw (2). After the set screw (2) is tightened, use a flat chisel or similar tool to stake screw in position.

MANUAL STEERING SYSTEM (460 INTERNATIONAL)

18. **ADJUSTMENTS.** The steering gear unit is provided with three adjustments which can be made without removing the steering gear assembly from the tractor: Cam (worm) shaft end play, the mesh position between the lever shaft stud and cam, and the end play of the gear shaft.

Before attempting to make any adjustments, first make certain that the gear housing is properly filled with lubricant, then disconnect the drag links from the steering (Pitman) arms.

19. CAM (WORM) SHAFT END PLAY. To check and/or adjust the steering camshaft end play, loosen both locks (8 and 8A—Fig. IH1017) and back-off the adjusting screws (9 and 9A) three or four turns. Pull up and push down on the steering wheel to detect any end play of the camshaft (27). Adjustment is correct when no end play exists and a barely perceptible drag is felt when turning the steering wheel with thumb and forefinger. If adjustment is not as specified, unbolt the upper cover (29) from gear housing and vary the number of shims (28) until desired adjustment is obtained. Shims are available in thicknesses of 0.002, 0.003 and 0.010. Tighten the adjusting screws (9 & 9A) as outlined in paragraphs 20 and 21.

20. LEVER SHAFT STUD MESH. With the camshaft end play adjusted as in paragraph 19, turn the steering gear to the mid or straight ahead position and tighten adjusting screw (9A—Fig. IH1017), located in side (trunion) cover (23) on right side of housing, until a slight drag is felt when turn-

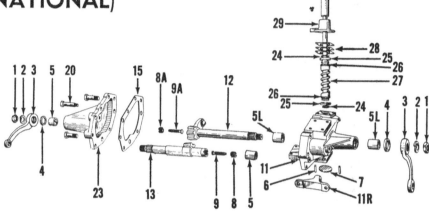

Fig. IH1017—Exploded view of the manual steering gear used on International 460 tractors. The lever shaft end play is adjusted by screw (9A) the gear shaft end play is adjusted by screw (9).

1. Nuts	6. Dowel pins
2. Lock washers	7. Expansion plug
3. Steering arms	8. Lock
4. Oil seals	8A. Lock
5. Gearshaft bushings	9. End play adjusting screw
5L. Levershaft bushings	9A. End play adjusting screw

11. Housing	27. Cam
12. Lever shaft	28. Shims
13. Gear shaft	29. Upper cover
15. Gasket	30. Jacket tube
23. Side cover	31. Bearing
24. Retaining rings	32. Spring seat
25. Ball cups	33. Spring
26. Steel balls	34. Dust seal

ing the steering gear through the mid or straight ahead position. Tighten the adjusting screw lock (8A).

21. GEAR SHAFT END PLAY. With the camshaft end play and the lever shaft mesh adjusted as outlined in paragraphs 19 and 20, turn the steering gear to the mid or straight ahead position and tighten adjusting screw (9—Fig. IH1017) to remove all end play from gear shaft (13) without increasing the amount of pull required to turn the steering gear through the mid or straight ahead position. Tighten the adjusting screw lock (8).

23. **REMOVE AND REINSTALL.** To remove the steering gear unit first drain cooling system and disconnect

drag links from the steering (Pitman) arms. Remove hood, battery and starting motor. Disconnect the heat indicator sending unit, fuel lines, oil pressure gage line, wiring harness and controls from engine and engine accessories. Disconnect head light and rear light wires. Unbolt steering gear housing and fuel tank from tractor and using a hoist, lift the fuel tank, instrument panel and steering gear housing assembly from tractor. NOTE: For convenience and ease of handling, some mechanics prefer to perform the slight additional work of removing the fuel tank before removing the steering gear.

24. Remove the steering wheel re-

taining nut and using a suitable puller, remove the steering wheel. Unbolt and remove the instrument panel assembly and fuel tank.

Note: Oversize dowels (6—Fig. IH-1017) are available if dowel holes have become worn. Holes in steering gear housing and clutch housing must be reamed to accommodate the oversize dowels.

25. **OVERHAUL.** To overhaul the steering gear, first remove the unit from tractor as outlined in paragraphs 23 and 24. Remove the steering (Pitman) arm retaining nuts (1—Fig. IH-1017) and using a suitable puller, remove the Pitman arms (3) from the lever shaft and gear shaft. Unbolt the side cover (23) from gear housing and remove the side cover and gear shaft (13). Withdraw lever shaft (12). Unbolt and remove the housing upper cover (29) and save shims (28) for reinstallation. Withdraw camshaft (27) and remove ball cups (25) by removing their retaining snap rings (24). Thoroughly clean and examine all parts for damage or wear. The lever

shaft and gear shaft should be renewed if the spur gear teeth are damaged or worn.

Inside diameter (new) of the lever shaft and gear shaft bushings is 1.3765-1.378. Diameter of lever shaft and gear shaft at bearing surfaces is 1.3725-1.3735. Renew the shafts and/or bushings if clearance is excessive.

New bushings are pre-sized and if carefully pressed into position using a suitable piloted arbor until outer ends of bushings are flush with inner edge of chamfered surface in bores, should need no final sizing.

When installing the camshaft and jacket tube, vary the number of shims (28) to remove all camshaft end play and provide a barely perceptible drag when turning steering wheel with the thumb and forefinger. Install the lever shaft, gear shaft and side cover. Install the steering (Pitman) arms.

Complete the overhaul by adjusting the lever stud mesh as in paragraph 20 and the gear shaft end play as outlined in paragraph 21.

adjusting screw (30—Fig. IH1020) until a very slight drag is felt only at the mid-point when turning steering wheel slowly from the full right to the full left position. Gear should rotate freely at all other points.

STEERING GEAR ASSEMBLY

29. **REMOVE AND REINSTALL.** To remove the steering gear assembly, first disconnect engine speed control and "Hydra-Touch" control valve linkage. Disconnect the drag link from Pitman arm. Remove the cap screws retaining gear housing to transmission housing cover and remove housing assembly from tractor.

30. **OVERHAUL.** To disassemble the removed steering gear assembly, mark the Pitman arm with respect to the lever shaft and remove the Pitman arm. Remove the housing side cover (29—Fig. IH1020) and withdraw cam lever and shaft. Remove the cap screws retaining steering column (9) to housing (17) and withdraw steering column, cam and bearing as a unit.

The camshaft lever stud assembly (24) is renewable.

When reassembling the gear unit, reverse the disassembly procedure and when installing the Pitman arm, be sure to align the previously affixed correlation marks. Adjust unit as outlined in paragraphs 27 and 28.

MANUAL STEERING SYSTEM
(560 AND 660 INTERNATIONAL)

ADJUSTMENTS. The steering gear is of the cam and lever type mounted on rear frame (transmission case) cover. Camshaft end play and gear backlash are adjustable, as follows:

27. CAM (WORM) SHAFT END PLAY. The cam (worm) shaft end play is controlled by shims (10—Fig. IH1020) located between steering tube (9) and steering gear housing (17). Adjustment is correct when camshaft (16) has zero end play and yet turns freely. To decrease end play remove shims. Shims are available in thicknesses of 0.002, 0.003 and 0.010. Disconnect drag link when making adjustment.

28. STEERING GEAR BACKLASH. Before adjusting the backlash, adjust the steering camshaft end play as in paragraph 27. Disconnect the drag link and place gear on the high point by turning steering wheel to mid-position of its rotation. Tighten cross shaft

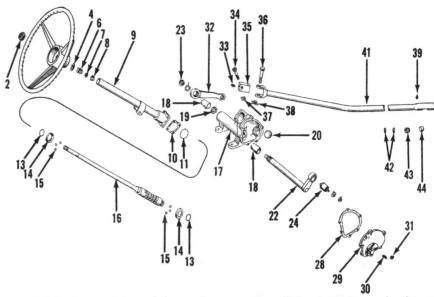

Fig. IH1020—Exploded view of the steering gear unit on 560 and 660 International tractors. Camshaft end play and backlash are adjustable.

4. Dust seal	11. Seal ring	19. Oil seal	30. Adjusting screw
6. Spring	13. Retaining rings	20. Expansion plug	31. Jam nut
7. Spring seat	14. Ball cups	22. Lever shaft	32. Pitman arm
8. Bearing	15. Bearing balls	23. Nut	35. Drag link
9. Jacket tube	16. Cam shaft	24. Stud	clevis block
10. Shims (0.002, 0.003 and 0.010)	17. Gear housing	28. Gasket	41. Drag link
	18. Bushings	29. Side cover	

POWER STEERING SYSTEM
(SERIES 460-560-660)

NOTE: The maintenance of absolute cleanliness of all parts is of utmost importance in the operation and servicing of the hydraulic power steering system. Of equal importance is the avoidance of nicks or burrs on any of the working parts.

LUBRICATION AND BLEEDING

All Models Without "Hydra-Touch"

35. The transmission and differential housing serves as the hydraulic fluid reservoir for the power steering system. The power steering fluid serves also as the lubricant for the "Torque Amplifier", transmission and differential gears. The fluid should be drained and renewed and the filter (Fig. IH1025) should be serviced at least every 1000 hours or once a year whichever occurs first.

NOTE: Early filters were of the wire screen type and had a by-pass valve (V—Fig. IH1026). Later filters were of the wire screen type but did not have the by-pass valve. The latest type (Fig. IH1026A) incorporates the wire screen and two renewable filter elements.

When either early type filter is encountered, it should be discarded and the latest type filter installed.

Only IH "Hy-Tran" fluid should be used in the hydraulic system and the reservoir fluid level should be maintained at the oil level plug located on the right side of the transmission and differential housing. Whenever the power steering oil lines have been disconnected or fluid drained, cycle the power steering system several times to bleed air from the system; then, refill the reservoir to the lower level of the oil level plug.

All Models With "Hydra-Touch"

36. On tractors with "Hydra-Touch", refer to paragraphs 265 and 266 concerning the lubrication and filter service requirements.

Fig. IH1025 — The power steering system filter should be cleaned in kerosene.

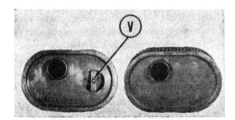

Fig. IH1026—Very early type filter elements were equipped with a by-pass valve (V). A later type filter element should be installed in place of the early type element. Refer to Fig. IH1026A for an exploded view of the latest type.

Fig. IH1027 — The power steering system operating pressure on series 460, 560 and 660 can be checked with a gage installed in place of plug (S).

OPERATING PRESSURE, RELIEF VALVE AND FLOW CONTROL VALVE

Series 460-560-660

37. Working fluid for the hydraulic power steering system is supplied by a pump contained in the clutch housing. The flow of approximately 2.3-2.7 gpm and pressure of 1200-1500 psi are regulated by spring loaded valves located in the pump body.

The overall condition of the system can be checked by installing a pressure gage of sufficient capacity in place of plug (S—Fig. IH1027). A gage reading of lower than 1100 psi with front wheels in either the extreme left or extreme right turn positions may be caused by one or more of the following: Clogged filter element; insufficient amount of oil; leaking lines; leaking seals; sticking or worn pressure relief valve; broken or weak relief valve spring; wear and consequent by-passing of oil in the power steering cylinder or control valve; worn or damaged pump.

To more accurately check the system and partially isolate the cause of trouble, proceed as follows: Connect a pressure gage of sufficient capacity and a shut-off valve in series with the pump pressure line (PL—Fig. IH1027). NOTE: The gage should be installed between the shut-off valve and the pump. Run tractor until power steering fluid reaches operating temperature, then close the shut-off valve and observe the gage reading. NOTE: If the shut-off valve is closed for an excessive length of time, damage to the pump may result. If the gage reading

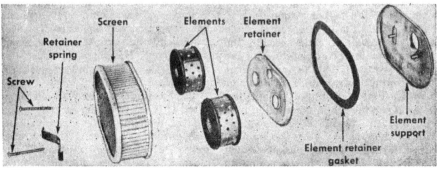

Fig. IH1026A—The latest type filter uses a wire screen and two renewable elements.

Screw — Retainer spring — Screen — Elements — Element retainer — Element support — Element retainer gasket

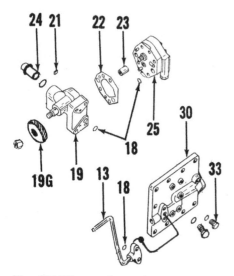

Fig. IH1030 — On early tractors not equipped with a hydraulic lift system pump the power steering pump (25) is mounted on adapter (19) and a drive coupling (23) is used.

13. Power steering	23. Coupling
pressure line	24. Suction tube
18. "O" rings	25. Power steering
19. Adapter	pump
19G. Drive gear	30. Pump mounting
21. Key	flange (manifold)
22. Gasket	33. Plugs

is 1200-1500 psi, the filter, pump, relief valve and pump intake seals are O.K. and any trouble is located in the pressure lines, power steering cylinder and/or control valve. Excessive pressure could be caused by the relief valve being stuck in the closed position.

If pressure is insufficient and the filter is known to be O.K., observe the oil. If the oil appears to be milky, an air leak in the suction line can be suspected. Air leakage at the pump intake seal can be corrected by removing the pump (or pumps) and mounting flange and renewing the intake pipe and seal (24—Fig. IH1031). If renewing the intake pipe and seal does not correct the condition after a reasonable length of time, the tractor should be split and seal (S—Fig. IH1164) renewed.

If pressure is low and the condition of the filter and the oil are known to be O.K., service the pressure relief valve.

The relief valve on both Cessna and Thompson pumps is pre-set and available only as an assembly. If cleaning the valve does not restore the pressure, the relief valve assembly (11R—Fig. IH1032) or (9—Fig. IH1036) should be renewed. Recheck pressure and if still insufficient, overhaul the pump.

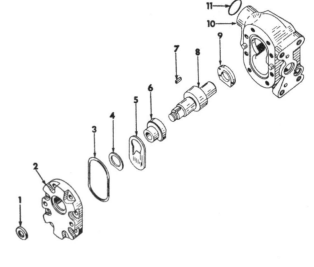

Fig. IH1030A—Exploded view of a late type pump adapter used when tractor is not equipped with hydraulic lift.

1. Oil seal
2. Cover
3. "O" ring
4. Spacer
5. Retainer plate
6. Bearing
7. Key
8. Adapter shaft
9. Bearing
10. Body
11. "O" ring

PUMP

Series 460-560-660

Cessna (Fig. IH1032) and Thompson (Fig. IH1036) power steering pumps are used interchangeably. Both pumps are of the gear type and are mounted on the rear of the hydraulic power lift pump or, on tractors without hydraulic power lift, on the rear of adapter (19—Fig. IH1030) also see Fig. IH-1030A.

38. REMOVE AND REINSTALL. The pump mounting flange with the pump (or pumps) attached can be removed from left side of tractor after the power steering pressure line and, if so equipped, the hydraulic lift system pressure line are detached from mounting flange and the flange retaining cap screws are removed. The removed flange and pumps are shown in Fig. IH1031.

The power steering pump can be separated from the hydraulic lift pump or drive bracket after removing the four attaching cap screws. Take care to prevent damage to the sealing surfaces when separating the power steering pump from the hydraulic pump or drive bracket. When reinstalling the power steering pump, tighten the four retaining screws and the two power steering pump cover to body screws evenly to 25 ft.-lbs. of torque.

When reinstalling the pumps and flange on tractor, make certain the longer smooth section of the suction pipe (24—Fig. IH1031) is pressed completely into the hydraulic lift pump or power steering pump drive adapter. The lip of the seal on the suction pipe should face toward the left (pump) side of the tractor. Vary the number and thicknesses of the gaskets which are located between pump mounting flange and clutch housing until the

Fig. IH1031 — View of the pumps and mounting flange. The longer smooth section of the suction tube (24) should be pressed in the hydraulic lift pump (or power steering pump drive adaptor) until it bottoms. Although Cessna pumps are shown, Thompson pumps are similar.

backlash between the pump driven gear and the PTO gear is 0.002-0.022. Gaskets are available in two thicknesses (0.011-0.013 and 0.021-0.023).

39. OVERHAUL CESSNA PUMP. To overhaul the removed pump, refer to Fig. IH1032 and proceed as follows: Remove the two pump body to pump cover cap screws, separate the cover from body and remove gears. Remove diaphragm (7), phenolic (back-up) gasket (6) and diaphragm seal (5) from cover (1).

The bushings are not available separately. The drive and driven gears, shafts and snap rings are available only as a complete set. Pressure relief valve (11R) can be disassembled and cleaned, but if its condition is questionable, renew the unit. Make certain

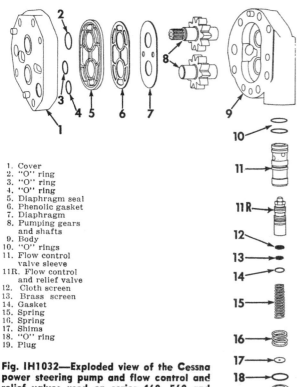

1. Cover
2. "O" ring
3. "O" ring
4. "O" ring
5. Diaphragm seal
6. Phenolic gasket
7. Diaphragm
8. Pumping gears
   and shafts
9. Body
10. "O" rings
11. Flow control
    valve sleeve
11R. Flow control
    and relief valve
12. Cloth screen
13. Brass screen
14. Gasket
15. Spring
16. Spring
17. Shims
18. "O" ring
19. Plug

Fig. IH1032—Exploded view of the Cessna power steering pump and flow control and relief valves used on series 460, 560 and 660. Shims (17) are used to adjust the volume of flow to 2.3-2.7 gpm.

1. Cover
2. "O" ring
3. "O" ring
4. Pumping gears and shafts
5. Pressure bearing assembly
6. "O" rings (2 used)
7. Retainers (2 used)
8. Pressure plate spring
9. Pressure relief valve
10. Gasket
11. Body
12. Flow control valve spring
13. Flow control valve
14. Plug
15. Snap ring

Fig. IH1036—Exploded view of the Thompson power steering pump and flow control and relief valves used on series 460, 560 and 660. The relief valve assembly is shown at (9). Some pumps may have plug (14) retained by a pin.

that valve (11R) slides freely in sleeve (11). Shims (17) control the pump output volume of 2.3-2.7 gpm. Be sure to install the same shims that were removed.

O.D. of shafts at bushings, min...0.685
I.D. of bushings in body
  and cover, max...............0.691
Thickness of gears, min.........0.457
I.D. of gear pockets in body,
  max.........................1.719

When reassembling, use a new diaphragm (7), phenolic gasket (6), diaphragm seal (5) and "O" rings (2, 3, 4, 10 and 18). With open part of diaphragm seal (5) towards cover (1), work same into grooves of cover using a dull tool. Press the phenolic gasket (6) into diaphragm seal (5) and install diaphragm (7) with bronze face towards gears. NOTE: The diaphragm must fit inside the raised rim of the diaphragm seal (5). Dip gears and shafts assemblies (8) in oil and install them in cover. Apply a thin coat of heavy grease on the finished side of pump body (9) and install body over gears and shafts. Check pump rotation. Pump should have very slight amount of drag but should rotate evenly.

40. OVERHAUL THOMPSON PUMP. Disassembly and overhaul procedures will be evident after an ex-

amination of the unit and reference to Fig. IH1036. All seals and questionable parts should be renewed. If any part of relief valve (9) appears questionable, renew the assembly. Make certain that flow control valve (13) slides freely in its bore.

Specifications are as follows:
O.D. of shafts at bushings, min...0.625
I.D. of bushings in body
  and cover, max.............0.6277
Thickness of gears, min........0.6865
I.D. of gear pockets in body,
  max. 1.449

STEERING CONTROL VALVE
Farmall 460-560

41. REMOVE AND REINSTALL. To remove the power steering control valve, detach the crankshaft pulley shield from the left side rail and the oil lines from the control valve. Remove the four Allen head screws which attach the control valve to the steering worm end cover.

When reinstalling the valve, renew the "O" ring which is located between the valve and the end cover. The valve actuating lever (29—Fig. IH1040) must be installed with larger end up. Make certain that the slots in the valve actuating lever properly engage the pin and the washer.

42. OVERHAUL. To overhaul the removed power steering control valve, refer to Fig. IH1040 and remove the two Allen head screws which retain the end cover (40) to valve body (45). Remove the self-locking nut (47N) from end of the clevis assembly (47). Remove the snap ring (50) and withdraw cover plate (49) and "O" ring (48). Remove parts (33, 34, 35, 36 and 37) from the rear end; then, withdraw the valve spool (46) and the clevis assembly (47) from the front.

NOTE: Do not attempt to remove the complete valve assembly from the rear because the clevis is slightly larger than the bore and damage to the bore may result.

A short length of small diameter wire inserted in the passage will facilitate removal of the valve assembly (41, 42, 43 and 44).

Renew all "O" rings and any parts which are in any way questionable. The valve spool and body are available only as a matched set.

Reassemble the control valve by reversing the disassembly procedure. The self-locking nut (47N) should be tightened until the washer (37) bottoms against the valve spool.

Series 460 International

43. R&R AND OVERHAUL. To remove the power steering valves, it is first necessary to remove the complete steering gear unit from tractor as outlined in paragraph 65.

With the gear unit removed from tractor, unbolt and remove the jacket tube and steering valve upper cover assembly. Refer to Fig. IH1043. Unbolt

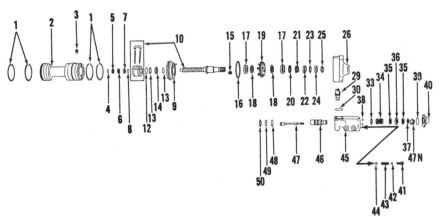

When reassembling the control valve, be sure to renew all "O" rings and seals and proceed as follows: Install the lower race (11L—Fig. IH1043) of the lower thrust bearing in the adapter casting (26). Pack the needle thrust bearing (12L) with grease and install bearing. Then, install upper race (14L) of lower thrust bearing. Press the two long inactive plungers (22) into the valve body with one plunger hole between them as shown in Fig. IH1045. Refer to Fig. IH1044 and install the six

Fig. IH1040—Exploded view of the Farmall 460 and 560 power steering worm, piston, control valve and associated parts. Twenty-one recirculating balls are used in the worm and nut assembly (10).

1. Rings	25. Oil seal
2. Piston and rack	26. End cover
3. Retaining screw	29. Valve actuating
4. Retaining ring	lever
5. Washer	30. "O" ring
6. Washer	33. Washer
7. Seal	34. Centering spring
8. Washer	35. Washers
9. Adapter	36. Seal ring
10. Worm and nut	37. Retainer
assembly	38. "O" rings
12. Retaining ring	39. "O" ring
13. Washers	40. Control valve end
14. Oil seal	cover
15. Centering springs	41. Plug and roll pin
(4 used)	42. "O" ring
16. "O" ring	43. Spring
17. Thrust bearing	44. Ball
races	45. Control valve body
18. Thrust bearings	46. Control valve
19. Center race and	spool
spring assembly	47. Clevis
20. Washer	47N. Nut
21. Lock nut	48. "O" ring
22. Oil seal	49. Cover plate
23. Washer	50. Snap ring
24. Bearing	

and remove the bearing adjusting nut (9) and lift out the upper thrust bearing (11, 12 and 14). Disconnect the oil lines from valve body and withdraw the valve assembly and lower thrust bearing. Be careful when removing the valve assembly and do not drop or nick any of the component parts.

Carefully slide the spool from the valve body and remove the six active plungers and the three centering springs as shown in Fig. IH1044. The two inactive plungers can be pressed from the valve body at this time. Thoroughly clean all parts in a suitable solvent and be sure all passages and bleed holes are open and clean.

The valve spool and body are mated parts and must be renewed as an assembly if damaged. Each of the centering springs should have a free length of 0.703 and should test 11-15 pounds when compressed to a height of 0.638 inches. The plungers and springs are available separately.

Fig. IH1043—Exploded view of 460 International power steering gear assembly. The adjusting screw (38L) controls the lever shaft stud mesh; screw (38G) controls gear shaft end play. The rack adjusting shims (45) are available in thicknesses of 0.003, 0.005 and 0.020.

4. Oil seal	18. Restricting plugs	20. Bushings (2 used)	44. Adjusting pad
5. Needle bearing	(9/64 O.D.)	30. Housing	45. Shims
7. Cover and tube	(2 used)	31. Bushings (4 used)	46. Cover
8. "O" rings	19. Control valve	32. Oil seals (2 used)	49. Piston rod and
9. Bearing adjusting	body	33. Steering arms	rack
nut	20. Check valve	34. Dowel pins	50. Back-up ring
11. Thrust bearing	21. Control valve	35. Expansion plug	51. "O" ring
race	spool	36. Bracket	52. Cylinder adapter
12. Thrust bearing	22. Inactive plungers	37. Lever shaft	53. "O" ring
13. "O" rings	(2 used)	38G. Adjusting screw	54. Piston
14. Thrust bearing	23. Active plungers	38L. Adjusting screw	55. Piston ring
race	(6 used)	39G. Retainer	56. Gasket
15. "O" rings	24. Spring (3 used)	39L. Retainer	57. Cylinder
16. Governor control	26. Valve adapter	40. Gear shaft	59. "O" rings
shaft pilot bushing	27. Cam (worm) and	41. Gasket	61. Front pipe
17. Restricting plug	shaft	42. Gasket	62. Rear pipe
(7/64 O.D.)	28. Gasket	43. Adjusting pad	
		plate	

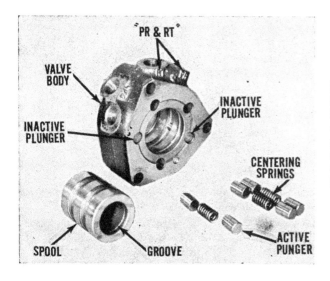

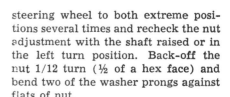

Fig. IH1044 — International 460 power steering control valve with the control spool, active plungers and centering springs removed. The inactive plungers can be be pressed from the valve body.

active plungers and the three centering springs in the remaining three plunger holes in valve body. Install the control spool (21—Fig. IH1043) in the valve body so that the identification groove in I.D. of spool is toward the same side of body as the port identification symbols "PR" and "RT". Refer to Fig. IH1046. Then install the assembled valve body with the symbols "PR" and "RT" up toward steering wheel. Install the upper thrust bearing assembly (11, 12 and 14—Fig. IH1043) with larger diameter race (14) toward valve body and be sure to pack the needle thrust bearing with grease. Install the tongued lock washer and nut (9), tighten the nut just enough to hold the assembly from slipping and center the upper thrust bearing lower race as follows:

Using a pair of dividers or six-inch scale, measure the distance from outer edge of the upper thrust bearing lower race (14) to outside edge of each of the five plunger holes. Shift the bearing race until these five measurements are identical. NOTE: The bearing race must be perfectly centered to prevent subsequent binding of the valve spool. Before tightening the adjusting nut, make certain that the lever shaft stud mesh, gear shaft end play and rack mesh are properly adjusted as in paragraphs 61, 62 and 64.

Install the steering wheel loosely on the shaft serrations and turn the steering wheel to the left, off the mid or straight ahead position, to lift the steering shaft to the upper extreme. Hold the steering wheel in this left turn position and tighten the valve adjusting nut to a torque of 10-12 ft.-lbs. If a suitable torque wrench is not available, a pair of 10-inch multi-slip joint pliers can be used. Turn the

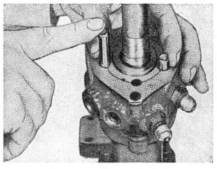

Fig. IH1045—Installing the inactive plungers in a 460 International power steering valve body. There must be one plunger hole (for active plungers) between the two inactive plungers as shown.

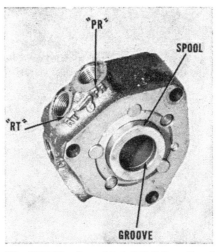

Fig. IH1046—Assembled 460 International power steering control valve. The spool should be installed so that groove in I.D. of same is toward same side of valve body as the cast in "PR & RT" markings.

Fig. IH1047 — Power steering cylinder and valve unit installation typical of that used on 560 and 660 International tractors.

steering wheel to both extreme positions several times and recheck the nut adjustment with the shaft raised or in the left turn position. Back-off the nut 1/12 turn (½ of a hex face) and bend two of the washer prongs against flats of nut.

Install the jacket tube assembly, mount a dial indicator and check the up and down movement of the steering worm (cam) shaft. When turning from the mid or straight ahead position to the extreme left, the shaft should move upward 0.050-0.055. When turning from the mid or straight ahead position to the extreme right, the shaft should move downward 0.050-0.055. In other words, when turning the steering wheel from one extreme position to the other, the shaft should have a total movement of 0.100-0.110. If the shaft movement is not as specified, it will be necessary to recheck the lever shaft stud mesh position and the valve nut adjustment.

The remainder of the assembly and reinstallation is reverse of the disassembly and removal procedures.

Series 560 and 660 International

44. **ADJUSTMENT.** To adjust the cylinder and valve unit, loosen the clamp on the steering drag link, then unstake and loosen the knurled adjusting plug lock nut. Refer to Fig. IH1047. Insert a ¼-inch rod in hole in the adjusting plug and turn the plug not more than ⅛-turn either way as required; then, road test tractor to determine if additional adjustment is needed.

NOTE: Turn the adjusting plug out to correct the following:

    a. Right turn is too hard
    b. Left turn is too easy
    c. Poor recovery after left turns
    d. Tractor wanders to the left

Turn the adjusting plug in to correct the following:

    a. Left turn is too hard
    b. Right turn is too easy
    c. Poor recovery after right turns
    d. Tractor wanders to the right

45. **OVERHAUL (EARLY).** With the cylinder and valve unit removed from tractor, mark the relative position of the drag link with respect to the valve adjusting plug so they can be reassembled in the same position, loosen the clamp and unscrew drag link from the adjusting plug. Remove

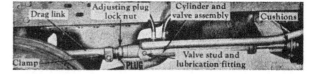

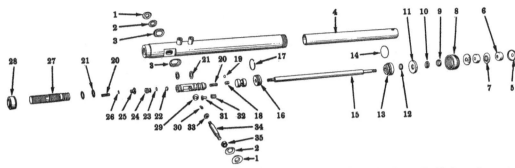

Fig. IH1048—Exploded view of early 560 and 660 International power steering cylinder and valve unit. If the valve spool, cylinder inner tube and/or housing are damaged, the complete unit should be renewed.

1. Stud cushion	7. Cushion locator	14. Guide seal ring	20. Reaction piston springs	25. Outer reaction piston	30. Lubrication fitting
2. Thrust washer	8. Cylinder tube cap	15. Piston rod		31. Inner ball seat
3. Cushion retainer	9. Wiper ring retainer	16. Piston	21. Valve seal rings	26. "O" ring	32. Spool spring
4. Cylinder inner tube	10. Wiper ring	17. Piston ring	22. Outer ball seat	27. Valve adjusting plug	33. Elastic stop nut
5. Cushion retainer	11. T-ring retainer	18. Inner reaction piston	23. Plug seal ring	28. Knurled lock nut	34. Valve stud
6. Frame bracket cushion	12. T-ring and washer	19. "O" ring	24. Spool adjusting plug	29. Valve stud ball	35. Slotted nut
	13. Cylinder rod guide			

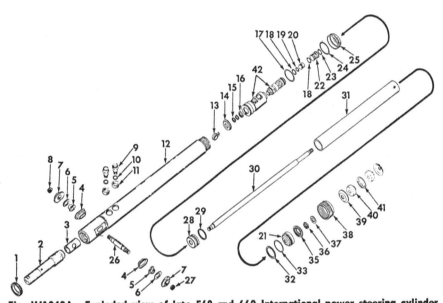

Fig. IH1048A—Exploded view of late 560 and 660 International power steering cylinder and valve unit. Valve body and spool (42) are available only as a matched unit.

1. Lock nut	11. Seal	22. Spool plug	33. "O" ring
2. Adjusting plug	12. Outer cylinder	23. "O" ring	35. T-ring & washer
3. Spacer	13. Spool rod end	24. "O" ring	36. Retainer
4. Cushion retainer	14. T-ring retainer	25. Cylinder base	37. Oil seal
5. Stud collar	15. T-ring washer	26. Valve stud	38. Cap
6. Cushion washer	16. T-ring	27. Nut (slotted)	39. Retainer
7. Stud cushion	17. "O" ring	28. Piston	40. Bracket cushion
8. Nut (plain)	18. "O" ring	29. Piston ring	41. Locator
9. Adapter	19. "O" ring	30. Piston rod	42. Valve body and spool
10. Gasket	20. Reaction piston	31. Inner cylinder	
	21. Piston rod guide	32. Retainer	

new all "O" ring seals and lubricate all parts in power steering fluid before installation. Install spool spring (32), inner ball seat (31), ball (29), outer ball seat (22) and adjusting plug (24). Turn the adjusting plug (24) completely in, then back the plug off ¼-turn and stake. Install the reactor springs (20) and reactor pistons (18 and 25). Install the assembled spool and screw the adjusting plug in to its original position. Install the knurled lock nut but do not tighten. Screw the drag link on the adjusting plug to its original position. Install the cylinder and valve unit on the tractor and adjust the unit as outlined in paragraph 44. Then tighten the drag link clamp and tighten and stake the knurled lock nut.

46. OVERHAUL (LATE). With the cylinder and valve unit removed from tractor, mark the relative position of the drag link with respect to the valve adjusting plug so they can be reassembled in the same position, loosen the clamp and unscrew drag link from the adjusting plug. Using a small chisel as shown in Fig. IH1049, unstake and remove the knurled lock nut; then, mark the depth to which the adjusting plug (2—Fig. IH1048A) is screwed into the housing (cylinder) and remove plug. Remove stud cushion (7), washer (6), collar (5) and retainer (4), pull stud (26) from cylinder assembly, then remove spacer (3). Remove adapter fittings (9), gasket (10) and seals (11). Remove cylinder tube cap (38) and guide assembly (21). Remove piston rod and piston assembly and inner tube (31). Cylinder tube base (25) and control valve assembly (42) can now be removed from outer tube (12). Drive out roll pin and separate rod end (13)

the through stud (34—Fig. IH1048) and withdraw the valve control arm. Using a small chisel as shown in Fig. IH1049, unstake and remove the knurled lock nut; then, mark the depth to which the adjusting plug is screwed into the housing and unscrew the plug from the housing. Using a large screw driver as shown in Fig. IH1050, push the valve assembly out of housing, but be careful not to scratch or otherwise damage any

of the polished surfaces. Pull out the reactor pistons (18 and 25—Fig. IH-1048) and remove the springs (20). Unscrew and remove plug (24) and disassemble the remaining parts. Thoroughly clean all parts in a suitable solvent and renew any which are damaged or show wear. If the valve spool or spool bore is damaged, the complete cylinder and valve unit should be renewed.

When reassembling, be sure to re-

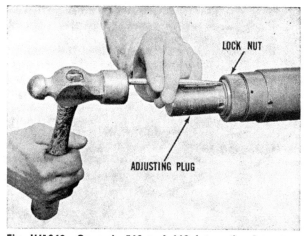

Fig. IH1049—On early 560 and 660 International tractors, the power steering valve adjusting plug can be unscrewed from body after unstaking and removing the knurled lock nut. Late cylinders will be similar.

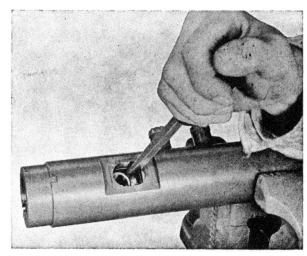

Fig. IH1050—On early 560 and 660 International tractors, the power steering valve can be pushed out of housing, using a large screw driver as shown.

from valve spool. Remove valve spool plug (22). Reaction piston (20) can now be removed from valve spool. Any further disassembly is obvious.

Clean and inspect all parts for wear, scoring or other damage and renew as necessary. Valve body and spool are available only as an assembly but all other parts are catalogued separately.

Use all new "O" rings, lubricate parts and reassemble by reversing the disassembly pocedure. Install the adjusting plug to its original position and install lock nut but do not tighten at this time. Screw drag link on the adjusting plug to its original position. Install the cylinder and valve unit on the tractor and adjust the unit as outlined in paragraph 44. Then tighten the drag link clamp and tighten and stake the knurled lock nut.

STEERING PISTON AND CYLINDER

Farmall 460-560

47. To remove the power steering piston, worm, actuator and end cover; proceed as follows: Remove any front end weights that may be attached to the steering gear housing; then, remove the sector shaft as outlined in paragraph 57. Remove the engine left side rail and the power steering control valve. Unbolt the steering shaft bracket from the clutch housing and pull the steering shaft front universal joint from the worm shaft. Unbolt the worm shaft end cover (26—Fig. IH-1055) from the gear housing and remove the end cover. Withdraw the piston, nut and worm shaft assembly.

If the steering cylinder sleeve needs to be removed, the four bolts that attach the gear housing to the right side

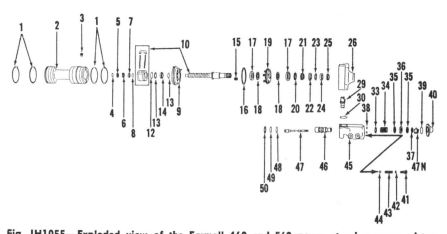

Fig. IH1055—Exploded view of the Farmall 460 and 560 power steering worm, piston, control valve and associated parts. Twenty-one recirculating balls are used in the worm and nut assembly (10.)

1. Rings	15. Centering springs (4 used)	26. End cover	41. Plug and roll pin
2. Piston and rack	16. "O" ring	29. Valve actuating lever	42. "O" ring
3. Retaining screw	17. Thrust bearing races	30. "O" ring	43. Spring
4. Retaining ring		33. Washer	44. Ball
5. Washer	18. Thrust bearings	34. Centering spring	45. Control valve body
6. Washer	19. Centering race and spring assembly	35. Washers	46. Control valve spool
7. Seal	20. Washer	36. Seal ring	47. Clevis
8. Washer	21. Lock nut	37. Retainer	47N. Nut
9. Adapter	22. Oil seal	38. "O" rings	48. "O" ring
10. Worm and nut assembly	23. Washer	39. "O" ring	49. Cover plate
11. Retaining ring	24. Bearing	40. Control valve end cover	50. Snap ring
12. Washers	25. Oil seal		
13. Washers			
14. Oil seal			

rail must be loosened enough to allow the gear housing to drop just enough to permit the withdrawal of the sleeve past the timing gear housing. When reinstalling the cylinder sleeve, make certain that the oil holes (H—Fig. IH-1056) in the sleeve and gear housing align.

To disassemble the piston, nut and worm shaft assembly; refer to Fig. IH-1055 and remove nut retaining screw (3). Withdraw the nut from the piston just slightly and wind a wire around the nut to prevent the ball guide from

falling out and losing the balls.

The need and procedure for further disassembly will be evident after an examination of the unit and reference to Fig. IH1055.

All "O" rings, seals, gaskets and questionable parts should be renewed.

If the nut (21) was removed, same should be tightened just enough to remove all end play of needle thrust bearings (18) without causing any binding tendency; then, the nut should be staked to maintain the correct adjustment.

Fig. IH1056 — The holes (H) in the sleeve and gear housing should be aligned when the sleeve is installed.

rack and adjusting pad for excessive wear. Be sure to renew all rings, seals and gaskets.

When reassembling, lubricate all parts and proceed as follows: Slide the piston rod through the cylinder end plate and install piston so that cupped end will be toward the retaining nut. Install the self-locking nut and tighten same securely as shown in Fig. IH1059. Install piston, end plate and rod assembly into cylinder. Check the lever stud mesh position as in paragraph 61

To reassemble, use a ring compressor to install the piston in the sleeve. Leave the piston protruding from the sleeve just far enough to install the worm shaft and nut assembly; then, secure the nut in the piston by installing the retaining screw (3) and bending part of screw into the groove in piston. Push piston the rest of the way in the sleeve and install end cover (26).

The remainder of the reassembly is the reverse for the disassembly procedure. After assembly is complete, start the engine and turn the steering wheel to the extreme right and the extreme left several times to bleed air from the system; then, check the fluid level in the transmission compartment.

Series 460 International

48. R&R AND OVERHAUL. To remove the power steering cylinder, remove the hydraulic lift system control valves rear and side covers. Disconnect the linkage and oil lines from the hydraulic system control valves and bracket. Unbolt the hydraulic control valves bracket from the clutch housing and lift same from tractor. Unbolt and remove the rack adjusting pad plate (43—Fig. IH1058) from bottom of gear housing. Disconnect oil lines from cylinder, then unbolt and remove cylinder and rack assembly from gear housing.

Using a plastic or lead hammer, bump the cylinder off the cylinder end plate and piston. Remove the nut retaining piston to piston rod, remove piston and withdraw the piston rod from the cylinder end plate. The procedure for further disassembly is evident. Thoroughly clean all parts in a suitable solvent and renew any damaged or worn parts. Be sure to examine rubbing surfaces of the gear

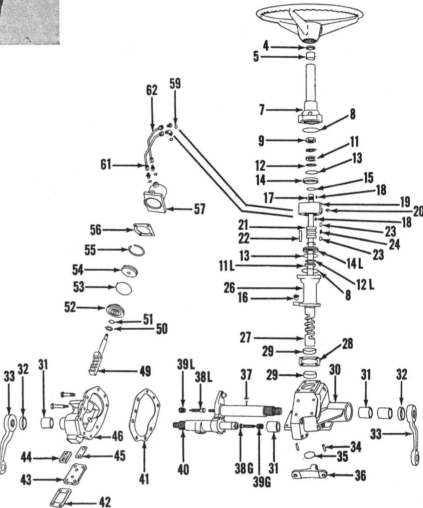

Fig. IH1058—Exploded view of 460 International power steering gear assembly. The adjusting screw (38L) controls the lever shaft stud mesh; screw (38G) controls gear shaft end play. The rack adjusting shims (45) are available in thicknesses of 0.003, 0.005 and 0.020.

4. Oil seal	18. Restricting plugs (9/64 O.D.) (2 used)
5. Needle bearing	19. Control valve body
7. Cover and tube	20. Check valve
8. "O" rings	21. Control valve spool
9. Bearing adjusting nut	22. Inactive plungers (2 used)
11. Thrust bearing race	23. Active plungers (6 used)
12. Thrust bearing	24. Springs (3 used)
13. "O" rings	26. Valve adapter
14. Thrust bearing race	27. Cam (worm) and shaft
15. "O" rings	28. Gasket
16. Governor control shaft pilot bushing	29. Bushings (2 used)
17. Restricting plug (7/64 O.D.)	

30. Housing	43. Adjusting pad plate
31. Bushings (4 used)	44. Adjusting pad
32. Oil seals (2 used)	45. Shims
33. Steering arms	46. Cover
34. Dowel pins	49. Piston rod and rack
35. Expansion plug	50. Back-up ring
36. Bracket	51. "O" ring
37. Lever shaft	52. Cylinder adapter
38G. Adjusting screw	53. "O" ring
38L. Adjusting screw	54. Piston
39G. Retainer	55. Piston ring
39L. Retainer	56. Gasket
40. Gear shaft	57. Cylinder
41. Gasket	59. "O" rings
42. Gasket	61. Front pipe
	62. Rear pipe

and the gear shaft end play as in paragraph 62, then install the assembled cylinder so that the first tooth of the gear sector meshes with the first space in the power rack.

Install the rack adjusting plate (43—Fig. IH1058) and vary the number of shims (45) between the pad (44) and the adjusting plate (43) to provide a slight drag between the rack teeth and gear shaft teeth when the pad retaining screws are securely tightened; then, deduct one 0.003 shim and reinstall the adjusting plate.

Series 560 and 660 International

49. OVERHAUL. With the cylinder and valve unit removed from tractor, unscrew the cylinder tube cap (8—Fig. IH1048) and pull piston and rod assembly from the cylinder tube. Thoroughly clean all parts and examine them for damage or wear.

NOTE: The cylinder is fitted with an inner tube (4) which is pressed into the rear valve retainer. If this inner tube is damaged, it is recommended that the complete cylinder and valve unit be renewed.

Piston rings should not bind in the piston grooves and rings should have an end gap of 0.001-0.006, when checked as shown in Fig. IH1060.

When reassembling, lubricate all parts in power steering fluid and renew all seals. Using a suitable ring compressor, install the piston and rod assembly as shown in Fig. IH1061. Tighten the cylinder end cap securely.

GEAR UNIT

Farmall 460-560

52. ADJUSTMENT. The steering gear unit is provided with two adjustments: The sector (upper bolster pivot) shaft end play and the worm shaft end play.

53. SECTOR SHAFT END PLAY. The sector (upper bolster pivot) shaft end play can be adjusted as follows: Remove the radiator as outlined in paragraph 175. Support the tractor in such a way as to remove weight from the front wheels. Remove the grille and the power steering oil line which provides oil to the sector shaft upper bearing. Remove the cap screws which retain the radiator bracket to the gear housing and lift bracket from tractor. NOTE: Four of the radiator bracket retaining cap screws can be removed from above the other three cap screws and one Allen head screw must be removed from below.

The adjusting nut (33—Fig. IH1062) should be tightened sufficiently to provide the bearings with a slight preload.

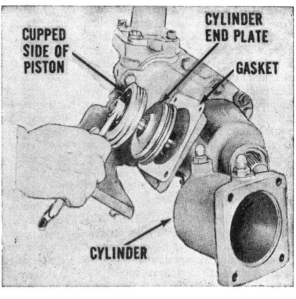

Fig. IH1059 — On 460 International tractors, the power steering piston should be installed with cupped side toward the retaining nut.

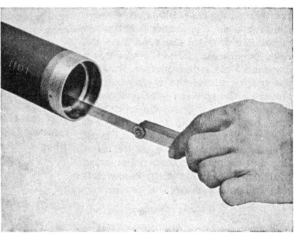

Fig. IH1060 — Checking the piston ring end gap in the 560 and 660 International power steering cylinder. Recommended end gap is 0.001-0.006. Early cylinder is shown, however, late cylinders will be similar.

Fig. IH1061 — Using a ring compressor to install 560 and 660 International power steering piston. All parts should be lubricated in power steering fluid prior to assembly. Early cylinder is shown, however, late cylinders will be similar.

54. WORM SHAFT END PLAY. The worm shaft end play can be adjusted as follows: Support the tractor in a suitable manner and remove the engine left side rail. Unbolt the steering shaft bracket from the clutch housing and slide the front universal joint from the worm shaft. Disconnect the oil lines from the power steering control valve and the worm shaft end cover (26—Fig. IH1055). Unbolt the control valve from the end cover and remove valve actuating lever (29). Unbolt the end cover from the steering

Fig. IH1062 — The upper bolster pivot shaft bearing adjusting nut (33) should be tightened sufficiently to provide the bearings with a slight pre-load.

gear housing and withdraw the cover from the worm shaft making certain that oil seals (22 and 25) are not damaged. Unstake nut (21) and tighten until all end play is just removed from the shaft without causing any binding tendency. NOTE: In most cases it will be necessary to renew the adjusting nut; also, the threads on the worm shaft should be inspected for damage. In some cases excessive end play may be due to the set screw (3) being loose in the worm shaft nut. Refer to paragraph 47 to remove the piston (rack) and tighten the set screw.

55. REMOVE AND REINSTALL. Removal of the complete steering gear and radiator assembly is necessary for jobs requiring removal of the crankshaft pulley, timing gear cover and others.

To remove the steering gear and radiator as an assembly, first remove the four hood side sheets (skirts) and unbolt the front of the hood from the radiator bracket. Drain coolant and disconnect the radiator hoses. Support the tractor under the clutch housing and sling the steering gear and radiator assembly in a hoist. On adjustable axle models, unbolt the radius rod pivot from the clutch housing. On all models, disconnect the pressure input line and return line from the steering control valve. Remove the engine left side rail and remove the cap screws

which retain the right side rail to the steering gear housing. Move the radiator and steering gear assembly forward away from the tractor.

When reinstalling, reverse the removal procedure. Make certain that the steering shaft front universal joint is guided on worm shaft as the radiator and steering assembly is rejoined to the tractor. Bleed the power steering system after reinstallation is complete and all lines are connected.

56. R&R AND OVERHAUL. Although most repair jobs associated with the steering gear will require the removal and overhaul of both the sector (upper bolster pivot) shaft and the worm shaft, there are some instances where failed or worn part is so located that repair can be accomplished by removing only one. In effecting such localized repairs, time will be saved by observing the following two paragraphs as a general guide.

57. SECTOR (UPPER BOLSTER PIVOT) SHAFT. The sector shaft (42 —Fig. IH1064) can be removed for service as follows: Remove the radiator as outlined in paragraph 175. Support the tractor in such a way as to remove the weight from the front wheels. Remove the grille and the

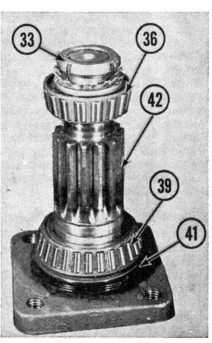

Fig. IH1065 — The upper bolster pivot shaft (42) and associated parts. Care must be taken to prevent damage to the seal (41).

Fig. IH1066 — When installing the upper bolster pivot shaft, the timing marks should be aligned as shown.

power steering oil line which provides oil to the sector shaft upper bearing. Remove the cap screws which retain the radiator bracket to the gear housing and lift bracket from tractor. NOTE: Four of the radiator bracket retaining cap screws can be removed from above; the other three cap screws and one Allen head screw must be removed from below. On adjustable axle equipped tractors, unbolt the radius rod pivot bracket from clutch housing. On all models, bend the locking tab from the adjusting nut (33—Fig. IH1062); then, remove the nut. Lift the front of the tractor up enough to clear the top of the sector shaft and roll wheel and sector shaft away from tractor.

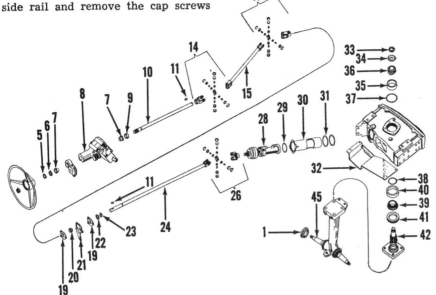

Fig. IH1064—Exploded view of Farmall 460 and 560 steering system. The worm unit (28) is shown exploded in Fig. IH1055.

1. Dust shield	15. Steering shaft (center)	26. Universal joint	36. Bearing cone
5. Snap ring	17. Universal joint	28. Worm unit	37. "O" ring
6. Washer	19. Bearing retainers	29. "O" ring	38. "O" ring
7. Bearings	20. Bearing	30. Sleeve	39. Bearing cone
8. Shaft housing	21. Bearing support bracket	31. "O" rings	40. Bearing cup
9. Shaft collar	22. Thrust washer	32. Crankshaft pulley shield	41. Oil seal
10. Steering shaft (rear)	23. Snap ring	33. Adjusting nut	42. Upper bolster pivot shaft
11. Woodruff keys	24. Steering shaft (front)	34. Lock washer	45. Dual wheel tricycle lower bolster
14. Universal joint		35. Bearing cup	

The lower bearing cone (39—Fig. IH1065) and seal (41) can be pulled from the sector shaft (42) and cups (35 and 40—Fig. IH1064) can be pulled from the gear housing if renewal is required.

When reassembling, renew all "O" rings and seals and mesh the line marked tooth on rack with valley between the two dotted teeth of sector gear. Refer to Fig. IH1066 for view of timing marks. The bearing adjusting nut should be tightened to provide bearings with a slight pre-load.

58. STEERING WORM UNIT. Refer to paragraph 47, for overhaul procedure for the power steering worm and piston assembly.

Series 460 International

60. **ADJUSTMENT.** The steering gear unit is provided with four adjustments: The mesh position between the lever shaft stud and camshaft, the end play of the gear shaft and the power rack mesh position can all be adjusted without removing the steering gear unit from tractor. The steering valve thrust bearings are adjusted by tightening the bearing adjusting nut to a torque of 10-12 ft.-lbs., but this should be done only when the valve unit is being serviced. Refer to paragraph 43.

Before attempting to make any adjustments, first make certain that the gear housing is properly filled with lubricant, then disconnect the drag links from the steering (Pitman) arms.

61. LEVER SHAFT STUD MESH. With the steering gear in the mid or straight ahead position, loosen the lock nut and back-off the gear shaft end play adjusting screw (38G—Fig. IH1067) three or four turns. Then loosen the lock nut and tighten the lever shaft adjusting screw (38L), located in cover on right side of housing, until a slight drag is felt when turning the steering gear through the mid or straight ahead position. Tighten the lever shaft adjusting screw lock nut.

62. GEAR SHAFT END PLAY. With the lever shaft stud mesh position adjusted as outlined in paragraph 61, turn the steering gear to the mid or straight ahead position and tighten adjusting screw (38G—Fig. IH1067) to remove all end play from gear shaft without increasing the amount of pull required to turn the steering gear through the mid or straight ahead position.

64. POWER RACK MESH. Refer to Fig. IH1067. With the rack adjusting plate (43) removed from the gear unit,

add shims (45) between pad (44) and plate (43) to provide a slight drag between the rack teeth and gear shaft teeth when the pad retaining screws are securely tightened; then, remove the adjusting plate and deduct one 0.003 shim and reinstall the adjusting plate.

65. **REMOVE AND REINSTALL.** To remove the steering gear unit first drain cooling system. Disconnect drag links from the steering (Pitman) arms; remove hood, battery and starting motor. Disconnect the heat indicator sending unit, fuel lines, oil pressure gage line, wiring harness and controls from engine and engine accessories. Disconnect head light and tail light wires. Disconnect the oil pressure and

return lines from the steering valves. Unbolt steering gear housing and fuel tank, from tractor and using a hoist, lift the fuel tank, instrument panel and steering gear housing assembly from tractor. NOTE: For convenience and ease of handling, some mechanics prefer to perform the slight additional work of removing the fuel tank before removing the steering gear.

66. Remove the steering wheel retaining nut and using a suitable puller, remove the steering wheel. Unbolt and remove the instrument panel assembly and fuel tank.

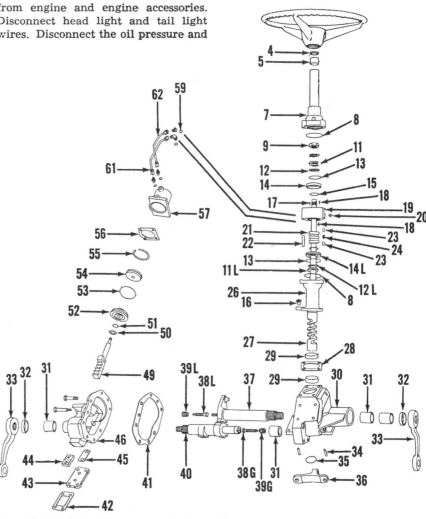

Fig. IH1067—Exploded view of 460 International power steering gear assembly. The adjusting screw (38L) controls the lever shaft stud mesh; screw (38G) controls gear shaft end play. The rack adjusting shims (45) are available in thicknesses of 0.003, 0.005 and 0.020.

4. Oil seal	18. Restricting plugs (9/64 O.D.) (2 used)	30. Housing
5. Needle bearing		31. Bushings (4 used)
7. Cover and tube	19. Control valve body	32. Oil seals (2 used)
8. "O" rings	20. Check valve	33. Steering arms
9. Bearing adjusting nut	21. Control valve spool	34. Dowel pins
	22. Inactive plungers (2 used)	35. Expansion plug
11. Thrust bearing race		36. Bracket
12. Thrust bearing	23. Active plungers (6 used)	37. Lever shaft
13. "O" rings		38G. Adjusting screw
14. Thrust bearing race	24. Springs (3 used)	38L. Adjusting screw
15. "O" rings	26. Valve adapter	39G. Retainer
16. Governor control shaft pilot bushing	27. Cam (Worm) and shaft	39L. Retainer
		40. Gear shaft
17. Restricting plug (7/64 O.D.)	28. Gasket	41. Gasket
	29. Bushings (2 used)	42. Gasket
		43. Adjusting pad plate

44. Adjusting pad	
45. Shims	
46. Cover	
49. Piston rod and rack	
50. Back-up ring	
51. "O" ring	
52. Cylinder adapter	
53. "O" ring	
54. Piston	
55. Piston ring	
56. Gasket	
57. Cylinder	
59. "O" rings	
61. Front pipe	
62. Rear pipe	

67. **OVERHAUL.** To overhaul the steering gear, first remove the unit from tractor as outlined in paragraphs 65 and 66. Remove the steering (Pitman) arm retaining nuts and using a suitable puller, remove the Pitman arms from the steering lever shaft and gear shaft. Refer to paragraph 43 for removal, reinstallation and overhaul of the steering valves and to paragraph 48 for R&R and overhaul of the steering cylinder.

With the valves and cylinder removed, unbolt the side cover from gear housing and remove the side cover and gear shaft. Withdraw lever shaft and camshaft. Thoroughly clean and examine all parts for damage or wear. The lever shaft and gear shaft should be renewed if the spur gear teeth are damaged or worn. Inspect also the roller bearing at upper end of the jacket tube.

Inside diameter (new) of the lever shaft and gear shaft bushings is 1.3765-1.378. Diameter of lever shaft and gear shaft at bearing surfaces is 1.3725-1.3735. Renew the shafts and/or bushings if running clearance is excessive. New bushings are pre-sized and if carefully pressed into position using a suitable piloted arbor until outer ends of bushings is flush with inner edge of chamfered surface in bores, should need no final sizing.

Inside diameter (new) of the camshaft bushings (29—Fig. IH1067) located in gear housing is 1.6235-1.6250. Diameter of camshaft at bushing surfaces is 1.620-1.621. Renew the camshaft and/or bushings if clearance is excessive. Bushings may need to be sized after installation to an inside diameter of 1.6235-1.6250.

Install the lever shaft and camshaft; then, install gear shaft in side cover. Install side cover and the steering (Pitman) arms.

Complete the gear unit overhaul by adjusting the lever stud mesh as in paragraph 61 and the gear shaft end play as outlined in paragraph 62. Then install the power cylinder and rack assembly and adjust the rack mesh position as in paragraph 64. Refer to paragraph 43 when installing the steering valves and adjusting the bearing nut.

Series 560 and 660 International

68. The procedures for adjusting, removing and reinstalling and overhauling the gear unit used on 560 and 660 International tractors which are equipped with power steering will be the same as similar procedures for manual steering. Refer to paragraphs 27, 28, 29 and 30.

POWER STEERING (SERIES 606 - 2606)

Series 606 and 2606 International tractors are available only with full time power steering. Working fluid for the full time (Hydrostatic) power steering is taken from the auxiliary hydraulic system pump via a flow divider valve. Basic components of the power steering system are: Hand pump, flow divider valve, pilot control valve, steering cylinder and oil cooler. Refer to Fig. IH1068 for a view showing the relative positions of the basic components.

LUBRICATION AND BLEEDING

International 606-2606

70. If steering cylinder has been emptied by servicing, or for any other reason, prime cylinder as follows: Disconnect the flexible cylinder lines at the aft end, hold lines up and slowly pour oil into both hoses. NOTE: Be sure to pour oil slowly in order to allow aid to escape. Reconnect lines as quickly as possible, then check to see that reservoir is at proper level. Start engine and run at low idle speed, then, rotate steering wheel (hand pump) as rapidly as possible in order to activate the control valve and continue to rotate the steering wheel until the front wheels reach the stop in the direction in which the steering wheel is being turned. Now quickly reverse the direction of steering wheel and follow same procedure until front wheels reach stop in opposite direction. Continue to turn front wheels from lock to lock until steering wheel has no wheel spin (free wheeling) and has a solid feel with no skips or sponginess. Check and add fluid to reservoir (rear frame), if necessary.

TROUBLE SHOOTING

International 606-2606

71. The following table lists some of the troubles, and their causes, which may occur in the operation of the series 606 and 2606 power steering system. When the following information is used in conjuction with the information contained in the Power Steering Operational Tests section (paragraph 73 through 79), no trouble should be encountered in locating system malfunctions.

1. No power steering or steers slowly.
   a. Binding mechanical linkage.
   b. Excessive load on front wheels and/or air pressure low in front tires.

c. Steering cylinder piston seal faulty or cylinder damaged.
   d. Control (pilot) valve relief valve setting too low or valve leaking.
   e. Faulty commutator in hand pump.
   f. Flow divider valve spool sticking or leaking excessively.
   g. Control (pilot) valve spool sticking or leaking excessively.
   h. Circulating check ball not seating.
2. Will not steer manually.
   a. Binding mechanical linkage.
   b. Excessive load on front wheels and/or air pressure low in front tires.
   c. Pumping element in hand pump faulty.
   d. Faulty seal on steering cylinder or cylinder damaged.
   e. Pressure check valve leaking.
   f. Control (pilot) valve spool binding or centering spring broken.
   g. Control (pilot) valve relief valve spring broken.
3. Hard steering through complete cycle.
   a. Low pressure from supply pump.
   b. Internal or external leakage.
   c. Line between hand pump and control (pilot) valve obstructed.
   d. Faulty steering cylinder.
   e. Binding mechanical linkage.
   f. Excessive load on front wheels and/or air pressure low in front tires.
   g. Cold hydraulic fluid.
4. Momentary hard or lumpy steering.
   a. Air in power steering circuit.
   b. Control (pilot) valve sticking.
5. Shimmy.
   a. Control (pilot) valve centering spring weak or broken.
   b. Control (pilot) valve centering spring washers bent, worn or broken.

OPERATING PRESSURE
International 606-2606

72. To check power steering operating pressure, refer to paragraphs 79 and 88, which give the method of checking the flow divider relief valve pressure.

POWER STEERING OPERATIONAL TESTS
International 606-2606

The following tests are valid only when the power steering system is completely void of any air. If necessary, bleed system as outlined in paragraph 70 before performing any operational tests.

73. MANUAL PUMP. With transmission pump inoperative (engine not running), attempt to steer manually in both directions. NOTE: Manual steering with transmission pump not running will require high steering effort. If manual steering can be accomplished with transmission pump inoperative, it can be assumed that the manual pump will operate satisfactorily with the transmission pump operating.

Refer also to paragraph 75 for information regarding steering wheel (manual pump) slip.

74. CONTROL (PILOT) VALVE. Attempt to steer manually (engine not running). Manual steering will require high steering effort but if steering can be accomplished, control (pilot) valve is working.

No steering can be accomplished if control valve is stuck on center. A control valve stuck off center will allow steering in one direction only. If excessive leakage of control valve is suspected, refer to section on Steering Wheel Slip for further checking.

75. STEERING WHEEL SLIP (CIRCUIT TEST). Steering wheel slip is the term used to describe the inability of the steering wheel to hold a given position without further steering movement. Wheel slip is generally due to leakage, either internal or external, or a faulty hand pump, steering cylinder or control (pilot) valve. Some steering wheel slip, with hydraulic fluid at operating temperature, is normal and permissible. A maximum of four revolutions per minute is acceptable. By using the steering wheel slip test and a process of elimination, a faulty unit in the power steering system can be located.

However, before making a steering wheel slip test to locate faulty components, quick disconnect assemblies should be installed in both steering cylinder lines and both hand pump to control (pilot) valve lines. This is necessary because of the fact that lines have to be disconnected and if it were not for the quick disconnects, air would be introduced into the circuit which in turn would not permit circuit testing having valid results. It is imperative that the power steering system be completely free of air before any testing is attempted.

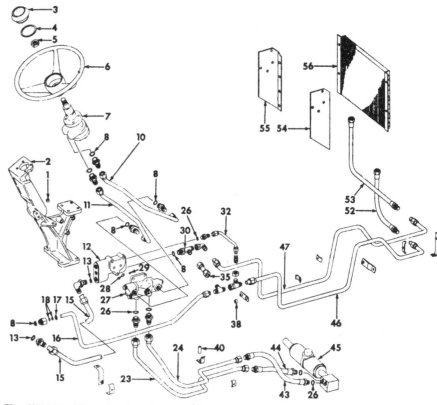

Fig. IH1068—View showing the general lay-out of components and tubing for the full time (Hydrostatic) power steering used on International 606 and 2606 tractors.

1. Governor control shaft bushing	15. Flow divider to transfer block tube	35. Oil cooler relief valve
2. Support	17. Retainer	38. Spacer
3. Steering wheel cap	18. "O" ring and back-up washer	40. Spacer
4. Cap retainer	23. Cylinder tube	43. Cylinder hose
5. Nut	24. Cylinder tube	44. Cylinder hose
7. Hand pump	26. "O" ring	45. Steering cylinder
8. "O" ring	27. Control (pilot) valve	46. Oil cooler return tube
10. Hand pump tube (LH)	28. "O" ring	47. Oil cooler inlet tube
11. Hand pump tube (RH)	29. "O" ring	52. Oil cooler outlet hose
12. Flow divider	30. Tee	53. Oil cooler inlet hose
13. "O" ring	32. Flow divider return tube	54. Oil cooler bracket (RH)
		55. Oil cooler bracket (LH)
		56. Oil cooler

To check for steering wheel slip (circuit test), proceed as follows: Check reservoir (rear frame) and fill to correct level, if necessary. Bleed power steering system, if necessary. Bring power steering fluid to operating temperature, cycle steering system until all components are approximately the same temperature and be sure this temperature is maintained throughout the tests. Remove steering wheel cap (monogram), then turn front wheels until they are against stop. Attach a torque wrench to steering wheel nut. NOTE: Either an inch-pound, or a foot-pound wrench may be used, however, an inch-pound wrench is recommended as it is easier to read. Advance hand throttle until engine reaches rated rpm, then apply 72 inch-pounds (6 foot-pounds) to torque wrench in the same direction as the front wheels are positioned against the stop. Keep this pressure

(torque) applied for a period of one minute and count the revolutions of the steering wheel. Use same procedure and check the steering wheel slip in the opposite direction. A maximum of four revolutions per minute in either direction is acceptable and system can be considered as operating satisfactorily. If, however, the steering wheel revolutions per minute exceed four, record the total rpm for use in checking the steering cylinder, control valve or hand pump.

NOTE: While four revolutions per minute of steering wheel slip is acceptable, it is generally considerably less in normal operation.

76. STEERING CYLINDER TEST. If steering wheel slip, as checked in paragraph 75, exceeds the maximum of four revolutions per minute, proceed as follows: Be sure operating temperature is being maintained, then disconnect the steering cylinder lines at quick disconnects. Repeat the steering wheel slip test, in both directions, as described in paragraph 75. If steering wheel slip is ½ rpm or more, less than that recorded in paragraph 75, overhaul or renew the steering cylinder.

If steering wheel slip remains the same as that obtained in the test outlined in paragraph 75, check the hand pump and control valve as outlined in paragraph 77.

77. MANUAL PUMP AND CONTROL VALVE. To check the hand pump and/or control valve, proceed as follows: Be sure system operating temperature is maintained, then disconnect the quick disconnects in the hand pump to control valve lines. Repeat the steering wheel slip test, in both directions, as described in paragraph 75.

If steering wheel slip is two rpms, or **more,** greater than those obtained in the circuit test outlined in paragraph 75, a faulty hand pump is indicated and hand pump should be serviced as outlined in paragraph 86.

If steering wheel slip is two rpms, or **less,** than those obtained in the steering cylinder test outlined in paragraph 76, a faulty control (pilot) valve is indicated and valve should be serviced at outlined in paragraph 87.

NOTE: At this time, the safety relief valve in the control (pilot) valve can be checked. To check this valve proceed as follows: Be sure hydraulic fluid is at operating temperature and hand pump to control valve lines are reconnected, then attach a gage, capable of registering at least 3000 psi, in series with one of the steering cylinder lines. Direct oil to the gage by turning the steering wheel in that direction until front wheels engage stop. With engine running at rated rpm, apply high steering effort to the steering wheel and observe the gage reading. Gage should read 2200 psi. If gage reading is not approximately 2200 psi, adjust safety relief valve by loosening jam nut and turning adjusting screw inward to increase pressure, or outward to decrease pressure. If safety relief valve will not adjust, remove and inspect spring, ball and seat. Renew spring and/or ball if necessary. If seat is damaged, renew complete control valve assembly as seat is not available separately. Recheck and readjust relief pressure. Relief valve adjusting screw is shown at (V—Fig. IH1069).

78. CONTROL (PILOT) VALVE TEST. Maintain the fluid temperature

Fig. IH1069—View of control (pilot) valve and flow divider valve mounting on International 606 and 2606 tractors.

C. Control valve
F. Flow divider valve
H. Hand pump lines
I. Oil cooler inlet line
O. Oil cooler relief valve
R. Return line
S. Steering cylinder lines
V. Safety relief valve

as outlined in paragraph 75, then turn front wheels against stop. Disconnect the control valve to steering cylinder line which is furnishing the pressurized oil in the direction of the turn. Repeat slip test of manual pump as outlined in paragraph 75. If slippage of steering wheel exceeds 2 rpm, or more, than that recorded in paragraph 75, service the hand pump as outlined in paragraph 86. If steering wheel slip is 2 rpm, or less, than that recorded in paragraph 75, the control valve is faulty and should be serviced as outlined in paragraph 87.

79. FLOW DIVIDER VALVE. The flow divider valve as applied to the International 606 and 2606 model tractors cannot be satisfactorily tested with unit installed in tractor. Therefore, to test the safety relief valve it is recommended that the unit be removed and bench tested. Refer to paragraph 88.

PUMP

International 606-2606

Cessna (Fig. IH1070) and Thompson (Fig. IH1070A) gear type pumps are used and are interchangeable. A priority flow of approximately 3 gpm is taken from the large pump (12 or 17 gpm) via a flow divider valve and the balance is utilized for the auxiliary hydraulic system, while the small (4½ or 7 gpm) pump supplies the draft and position control system if tractor is so equipped. Also see Fig. IH1070D.

80. REMOVE AND REINSTALL. The pump mounting flange (Fig. IH1070B) with the pump (or pumps) attached can be removed from left side of tractor after the power steering pressure line and the hydraulic lift system pressure line are detached

from mounting flange and the flange retaining cap screws are removed. The removed flange and pumps are shown in Fig. IH1070C.

NOTE: If tractor is not equipped with draft and position control, the large pump will be fitted with a cover in place of small pump.

The small pump can be separated from the large pump after removing the four attaching cap screws. Take care to prevent damage to the sealing surfaces when separating the pumps. When reinstalling the small pump, tighten the four retaining screws and the two pump covers to body screws evenly to 25 Ft.-Lbs. of torque.

When reinstalling the pumps and flange on tractor, make certain the longer smooth section of the suction pipe (24—Fig. IH1070C) is pressed completely into the large pump. The lip of the seal on the suction pipe should face toward the left (pump) side of the tractor. Vary the number and thickness of the gaskets which are located between pump mounting flange and clutch housing until the backlash between the pump driven gear and the PTO gear is 0.002-0.026. Gaskets are available in two thicknesses; 0.011-0.019 and 0.016-0.024.

NOTE: On series 606 and 2606 tractors prior to serial number 503, a 14 tooth pump drive gear (11—Fig. IH1070D) was used. On tractors serial number 503 to 1016 a 15 tooth pump drive gear was used. On tractors serial number 1016 to 2897, a 17 tooth drive gear was used. If a 17 tooth gear is being installed on a tractor serial number 503 to 1016, the original flange (19) can be used providing a spacer (18) is also installed. Spacer is used to compensate for the increased diameter of the 17 tooth pump drive gear. Tractors serial number 2897 and up use a 15 tooth gear.

81. OVERHAUL CESSNA PUMP.
To overhaul the removed pump, remove pump flange and pumps as outlined in paragraph 80, then refer to Fig. IH1070 and proceed as follows: Remove the four retaining cap screws and separate the small pump from large pump, then remove large pump from pump mounting flange.

NOTE: If tractor is not equipped with draft and position control, cover (14) and gasket (13) will be used on pump body in place of the small pump.

Remove pump drive gear, then unbolt and remove pump cover (2). The remainder of disassembly will be obvious after an examination of the unit and reference to Fig. IH1070.

Bushings in pump body and cover are not available separately and the pump gears and shafts are available only in sets. Gaskets and "O" rings are also available in a package only.

Pump specifications are as follows:

12 GPM Pump
O. D. of shafts at bearings (min.) 0.810
I. D. of bushings in body
  and cover (max.) 0.816
Thickness (width) of gears
  (min.) 0.572
I. D. of gear pockets (max.) 2.002
Max. allowable shaft to
  bushing clearance 0.006

17 GPM Pump
O. D. of shafts at bearings (min.) 0.810
I. D. of bushings in body
  and cover (max.) 0.816
Thickness (width) of gears
  (min.) 0.813
I. D. of gear pockets (max.) 2.002
Max. allowable shaft to
  bushing clearance 0.006

When reassembling, use new diaphragms, gaskets, back-up washers, diaphragms seal and "O" rings. With open part of diaphragm seal (5) towards cover (2), work same into grooves of cover using a dull tool. Press protector gasket (6) and back-up gasket (7) into the relief of diaphragm seal. Install check ball (3) and spring (4) in cover, then install diaphragm (8) inside the raised lip of the diaphragm seal and be sure bronze face of diaphragm is toward pump gears. Dip gear and shaft assemblies in oil and install them in cover. Position diaphragm (15) in pump body with the bronze side toward pump gears and cut-out portion toward inlet (suction) side of pump. Install pump body over gears and shafts and install retaining cap screws. Torque cap screws to 20 ft.-lbs. for the 12 gpm pump and to 25 ft.-lbs. for the 17 gpm pump.

Check pump rotation. Pump will

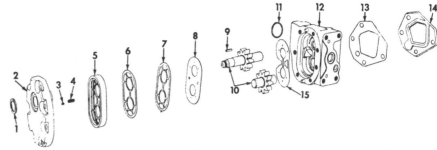

Fig. IH1070—Exploded view of the Cessna pump used for the power steering system and auxiliary hydraulic system of the International 606 and 2606 tractors.

1. Oil seal	9. Key
2. Cover	10. Pump gears and
3. Check ball	shafts
4. Spring	11. "O" ring
5. Diaphragm seal	12. Pump body
6. Protector gasket	13. Gasket
7. Back-up gasket	14. Rear cover
8. Pressure diaphragm	15. Body diaphragm

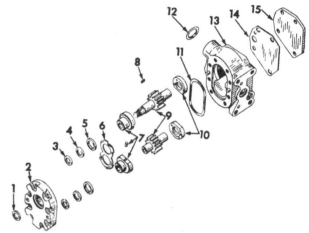

Fig. IH1070A—Exploded view of the Thompson pump used for the power steering system and auxiliary hydraulic system of the International 606 and 2606 tractors.

1. Oil seal
2. Cover
3. Back-up washer
4. "O" ring
5. Retainer
6. Pressure plate spring
7. Pressure bearings
8. Key
9. Pump gears and shafts
10. Body bearings
11. "O" ring
12. "O" ring
13. Body
14. Gasket
15. Rear cover

Fig. IH1070B—View of pump mounting flange. Small line feeds draft and position control. Large line feeds power steering and auxiliary systems via a flow divider valve.

have a slight amount of drag but should turn evenly.

82. OVERHAUL THOMPSON PUMP. To overhaul the removed pump, remove pump flange and pumps as outlined in paragraph 80, then refer to Fig. IH1070A and proceed as follows: Remove the four retaining cap screws and separate small pump from large pump, then remove large pump from pump mounting flange.

NOTE: If tractor is not equipped with draft and position control, cover (15) and gasket (14) will be used on pump body in place of the small pump. Remove pump drive gear and key (8), then unbolt and remove pump cover (2). Bearings (7), pressure plate spring (6), "O" ring retainers (5), "O" rings (4), back-up washers (3) and oil seal (1) can now be removed from cover. Note location of bearings (7) so they can be reinstalled in the

Fig. IH1070C—View of the pumps and mounting flange. The longer smooth section of the suction tube should be pressed in the hydraulic lift pump (or power steering pump drive adapter) until it bottoms. Although Cessna pumps are shown, Thompson pumps are similar.

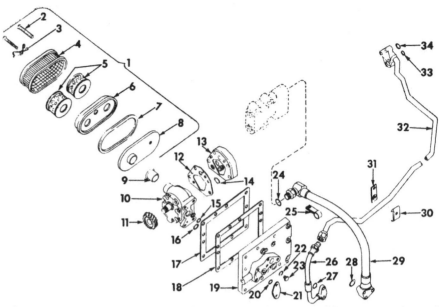

Fig. IH1070D—On series 606 and 2606 International tractors, with power steering and draft and position control, two pumps are used as shown. On models with no draft and position control, small pump (13) is not used and a cover is installed on large pump (10).

same position. Remove "O" rings (11 and 12) and the pump gears and shafts (9) from pump body. Bump body on a wood block and remove bearings (10). Note location of bearings so they can be reinstalled in the same position.

Pump gears and shafts, as well as the pump shaft bearings, are available only in sets. Except for suction port "O" rings (12), none of the gaskets or "O" rings are available separately.

Pump specifications are as follows:

12 GPM Pump

O. D. of shafts at bearings (min.)	0.812
I. D. of bearings in body and cover (max.)	0.816
Thickness (width) of gears (min.)	0.7765
I. D. of gear pockets (max.)	1.772
Max. allowable shaft to bearing clearance	0.004

17 GPM Pump

O. D. of shafts at bearings (min.)	0.812
I. D. of bearings in body and cover (max.)	0.816
Thickness (width) of gears (min.)	1.072
I. D. of gear pockets (max.)	1.772
Max. allowable shaft to bearing clearance	0.004

Lubricate all parts during assembly, use all new gaskets and seals and be sure bearings in body and cover are reinstalled in their original positions if same bearings are being reinstalled. Tighten cover to body cap screws to a torque of 20 ft.-lbs. for the 12 gpm pump, or 30 ft.-lbs. for the 17 gpm pump.

Check pump rotation. Pump will have a slight amount of drag but should turn evenly.

Fig. IH1071—Oil cooler for the power steering system of series 606 and 2606 International tractors is mounted as shown.

OIL COOLER AND RELIEF VALVE

International 606-2606

83. The oil cooler is mounted in front of radiator as shown in Fig. IH1071. This unit is used to cool only the oil used by the power steering system. Also incorporated into the oil cooler system is a relief (by-pass) valve (O-Fig. IH1069) which opens and by-passes the oil to reservoir, should the oil cooler become plugged or if oil is too cold to circulate.

Oil cooler can be unbolted from radiator after removing hood and grille and disconnecting hoses. Relief valve can be removed at any time and procedure for doing so is obvious.

Relief valve can be bench tested and should open at 85 psi. Correct faulty units by renewing same.

STEERING CYLINDER

International 606-2606

84. **R & R AND OVERHAUL STEERING CYLINDER.** To remove the power steering cylinder, disconnect and immediately plug the hydraulic lines. Remove cap screws from the steering cylinder support (steering arm) and disengage from steering cylinder. Remove pin retaining anchor assembly to axle and remove cylinder from tractor.

With cylinder removed, move piston rod back and forth several times to clear oil from cylinder. Refer to Fig. IH1072 and proceed as follows: Place end of piston rod which has the flats in a vise, then unscrew and remove anchor assembly. (15). Remove cylinder head retainer (7) as follows: Lift end of retainer out of slot, then using a pin type spanner, rotate cylinder head (2) and work retainer out of its groove. Cylinder head and the piston rod and piston assembly (10) can now be removed from cylinder. Remove the remaining cylinder head in the same manner. All seals, "O" rings and back-up washers are now available for inspection and/or renewal.

Clean all parts in a suitable solvent and inspect. Check cylinder for scoring, grooving and out-of-roundness. Light scoring can be polished out by

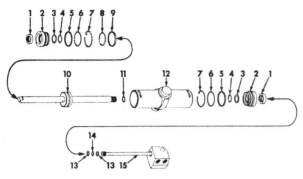

Fig. IH1072—Exploded view of the power steering cylinder used on series 606 and 2606 International tractors.

1. Wiper seal
2. Cylinder head
3. Back-up washer
4. "O" ring
5. Back-up washer
6. "O" ring
7. Cylinder head retainer
8. Piston seal "O" ring
9. Piston seal
10. Piston and rod
11. "O" ring
12. Cylinder
13. Back-up washer
14. "O" ring
15. Anchor assembly

using a fine emery cloth and oil providing a rotary motion is used during the polishing operation. A cylinder that is heavily scored or grooved, or that is out-of-round, should be renewed. Check piston rod and cylinder for scoring, grooving and straightness. Polish out very light scoring with fine emery cloth and oil, using a rotary motion. Renew rod and piston assembly if heavily scored or grooved, or if piston rod is bent. Inspect piston seal (9) for frayed edges, wear and imbedded dirt or foreign particles. Renew seal if any of the above conditions are found. NOTE: Do not remove the "O" ring (8) located under the piston seal unless renewal is indicated as it is not necessary to renew this "O" ring unless it is damaged. Inspect balance of "O" rings, back-up washers and seals and renew as necessary. Inspect bores of cylinder heads and renew same if excessively worn or are out-of-round.

Reassemble steering cylinder as follows: Place "O" ring (14), with back-up washer (13) on each side, in groove at inner end of anchor assembly oil tube. Install piston rod "O" ring (11) in piston rod hole in anchor assembly. Install wiper seal (1), back-up washer (3), piston rod "O" ring (4), cylinder head "O" ring (6) and back-up washer (5) to cylinder head, then install the cylinder head assembly over threaded end of piston rod. Lubricate "O" ring and back-up washers on inner end of anchor assembly oil tube and carefully insert into threaded end of piston rod. Lubricate the piston rod "O" ring (11) and push anchor assembly toward piston rod. As "O" ring on inner end of oil tube approaches the drilled hole (port) in the piston rod (located near piston), use IHC tool FES 65, or equivalent, to depress "O" ring and washers so they will pass the port without being damaged. Screw anchor assembly onto

piston rod and tighten to a torque of 150 ft.-lbs. Lubricate piston seal and cylinder head "O" rings and using a ring compressor, or a suitable hose clamp, install piston and rod assembly into cylinder. Install cylinder head in cylinder so hole in cylinder head will accept nib of retaining ring, then position retaining ring and pull same into its groove by rotating cylinder head. Complete balance of re-assembly by reversing disassembly procedure.

Reinstall unit on tractor, then fill and bleed the power steering system as outlined in paragraph 70.

NOTE: Prior to installing steering cylinder on tractor, inspect the bushings in the upper and lower cylinder supports (steering arms) and renew if necessary. Bushings are available separately and renewal procedure is obvious.

HAND PUMP

International 606-2606

85. REMOVE AND REINSTALL. To remove the hand pump used in International 606 and 2606 tractors, remove hood, and steering wheel. NOTE: Use a puller to remove steering wheel, do not bump on upper end of steering wheel shaft. Remove throttle lever. Unbolt instrument panel from instrument panel housing (cowl) and remove housing. Disconnect tachometer cable and move instrument panel out of the way. Remove battery top cover. Disconnect lines from hand pump, then unbolt and remove hand pump from steering shaft support.

Reinstall by reversing the removal procedure and bleed power steering system as outlined in paragraph 70.

86. OVERHAUL MANUAL (HAND) PUMP. Remove the manual pump as outlined in paragraph 85. Clear fluid from unit by rotating steering wheel

(input) shaft back and forth several times. Place unit in a soft jawed vise with end plate on top side, then remove end plate retaining cap screws and lift off end plate (2—Fig. IH1073).

NOTE: Lapped surfaces of end plate (2), pumping element (5), spacer (6), commutator (9) and pump body (13) must be protected from scratching, burring or any other damage as sealing of these parts depends only on their finish and flatness.

Remove seal retainer (3), seal (4), pumping element (5) and spacer (6) from body (13). Remove commutator (9) and drive link (8), with link pins (7) and commutator pin (9A), from body. Smooth any burrs or nicks which may be present on input shaft (10), wrap spline with masking tape, then remove input shaft from body. Remove bearing race (12) and thrust bearings (11) from input shaft. Remove snap ring (19), washer (18), spacer (17), back-up washer (16) and seal (15). Do not remove needle bearing (14) unless renewal is required. If it should be necessary to renew bearing, press same out pumping element end of body.

Clean all parts in a suitable solvent and if necessary, remove paint from outer edges of body, spacer and end plate by passing these parts lightly over crocus cloth placed on a perfectly flat surface. Do not attempt to dress out any scratches or other defects since these sealing surfaces are lapped to within 0.0002 of being flat. However, in cases of emergency, a spacer that is damaged on one side only may be used if the smooth side is positioned next to the pumping element and the damaged side is lapped flat.

Inspect commutator and housing for scoring and undue wear. Bear in mind that burnish marks may show, or discolorations from oil residue may be present, on commutator after unit has been in service for some time. These can be ignored providing they do not interfere with free rotation of commutator in body.

Check fit of commutator pin in the commutator. Pin should be a snug fit and if bent, or worn until diameter at contacting points is less than 0.2485, renew pin.

Measure inside diameter of input shaft bore in body and outside diameter of input shaft bearing surface. If body bore is 0.006, or more, larger than shaft diameter, renew shaft and/or body and commutator. Note: Body and commutator are not available separately.

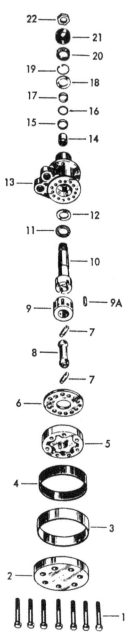

1. Cap screw	11. Thrust bearing
2. End plate	12. Bearing race
3. Seal retainer	13. Body
4. Seal	14. Needle bearing
5. Pumping element	15. Seal
6. Spacer plate	16. Back-up washer
7. Link pin	17. Spacer
8. Drive link	18. Washer
9. Commutator	19. Retainer
9A. Commutator pin	20. Felt seal
10. Coupling (input) shaft	21. Water seal
	22. Nut

Check thrust bearing and race for excessive grooving, flat spots or any other damage and renew bearing assembly if necessary.

Place pumping element on a flat surface and in the position shown in Fig. IH1073A. Use a feeler gage and check clearance between ends of ro-

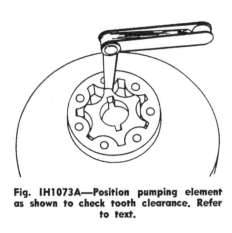

Fig. IH1073A—Position pumping element as shown to check tooth clearance. Refer to text.

tor teeth and high points of stator, If clearance exceeds 0.003, renew pumping element. Use a micrometer and measure width (thickness) of rotor and stator. If stator is 0.002 or more wider (thicker) than the rotor, renew the pumping element. Pumping element rotor and stator are available only as a matched set.

Check end plate for wear, scoring and flatness. Do not confuse the polish pattern on end plate with wear. This pattern, which results from rotor rotation, is normal. Renew end plate if worn or scored and is not within 0.00002 of being flat.

When reassembling, use all new seals and back-up washers. All parts, except those noted below, are installed dry. Reassemble as follows: If needle bearing (14—Fig. IH1073) was removed, lubricate with IH Hy-Tran fluid, install from pumping element end of body and press bearing into bore until inside end measures $3\frac{13}{16}$-3⅞ inches from pumping element end of body as shown in Fig. IH1073B. Lubricate thrust bearings assembly with IH Hy-Tran fluid and install assembly on input shaft with race on top side. Install input shaft and bearing assembly in body and check for free rotation. Install a link pin in one end of the drive link, then install drive link in input shaft by engaging the flats on link pin with slots in input shaft. Use a small amount of grease to hold commutator pin in commutator, then install commutator and pin in body while engaging pin in one of the long slots of the input shaft. Commutator is correctly installed when edge of commutator is slightly below sealing surface of body. Clamp body in a soft jawed vise with input shaft pointing downward. Again make sure surfaces of spacer, pumping element, body and end plate are perfectly clean, dry and undamaged. Place spacer on body and align screw holes with those of body. Put link pin in

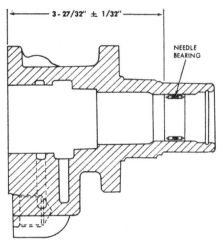

Fig. IH1073B—When renewing needle bearing in body, install same to dimension shown.

exposed end of drive link, then install pumping element rotor while engaging flats of link pin with slots in rotor. Position pumping element stator over rotor and align screw holes of stator with those of spacer and body. Lubricate pumping element seal lightly with IH Hy-Tran fluid and install seal in seal retainer, then install seal and retainer over pumping element stator. Install end cap, align screw holes of end cap with those in pumping element, spacer and body, then install cap screws. Tighten cap screws evenly and to a torque of 18-22 ft.-lbs.

NOTE: If input shaft does not turn evenly after cap screws are tightened, loosen and retighten them again. However, bear in mind that the unit was assembled dry and some drag is normal. If stickiness or binding cannot be eliminated, disassemble unit and check for foreign material, nicks or burrs which could be causing interference.

Lubricate input shaft seal with IH Hy-Tran fluid and with input shaft splines taped to protect seal, install seal, back-up washer, spacer, washer and snap ring. The felt washer and water seal may be installed at this time but there will be less chance of loss or damage if installation is postponed until the time the steering wheel is installed.

After unit is assembled, turn unit on side with hose ports upward. Pour until full of oil and work pump slowly until interior (pumping element) is thoroughly coated. Either plug ports or drain excess oil.

Reinstall unit by reversing the removal procedure and bleed steering system as outlined in paragraph 70.

CONTROL (PILOT) VALVE

International 606-2606

87. **R&R AND OVERHAUL.** Procedure for removal of control valve is obvious after an examination of the unit and reference to Fig. IH1069.

With valve removed, disassemble as follows: Refer to Fig. IH1074 and remove end caps (1) with "O" rings (2). Pull spool and centering spring assembly from valve body. Place a punch or small rod in hole of centering spring screw (3) and remove screw, centering spring (5) and centering spring washers (4) from spool. Remove plug (11), "O" ring (12) and circulating check ball (10). Remove retainer (16), seat (15), pressure check valve (14) and spring (13).

Fig. IH1074A—Early type flow divider valve with component parts removed.

　　A. Adapter
　　B. Valve body
　　P. Plug
　　R. Safety (relief) valve
　　S. Spring
　　V. Valve spool

Do not disassemble safety relief valve (items 6 through 10) unless it is deemed absolutely necessary. Valve is set at factory and in normal operation, is seldom actuated. Therefore, it is unlikely that valve will be damaged. However, should it be necessary to disassemble valve, count and record the number of exposed threads on the adjusting screw and be sure to reinstall the adjusting screw to the same position.

Wash all parts in a suitable solvent and inspect. Valve spool and spool bore in body should be free of scratches, scoring or excessive wear. Spool should fit its bore with a snug fit and yet move freely with no visible side play. If spool or spool bore is defective, renew complete valve assembly as spool and valve body are not available separately.

Inspect pressure check valve and seat. Renew parts if grooved or scored.

If safety relief valve was disassembled, inspect spring for fractures, distortion or lack of tension. Inspect ball for nicks and grooves. Renew parts as necessary. If seat is damaged, it will be necessary to renew complete valve as seat is not available separately. However, very light scoring or nicks can be removed from seat by using a new ball, with fine grinding compound, to lap the seat. Be absolutely certain that valve body is completely cleaned and all traces of grinding compound removed.

Reassembly is the reverse of disassembly and the following points should be observed. Coat all parts with Hy-Tran fluid, or its equivalent, prior to installation. If safety relief valve was disassembled, be sure the adjusting screw is installed with the same number of threads exposed as were exposed prior to removal. Measure distance between gasket surface of circulating check ball plug and inner end of roll pin. This distance should be $\frac{15}{16}$-inch and if necessary, obtain this measurement by adjusting roll pin in or out. Tighten end cap retaining cap screws to a torque of 22-27 ft.-lbs.

Reinstall valve by reversing removal procedure and bleed power steering

system as outlined in paragraph 70.

NOTE: If necessary, the pressure setting of the safety relief valve can be checked and/or adjusted as outlined in paragraph 77.

FLOW DIVIDER VALVE

International 606-2606

88. **R&R AND OVERHAUL.** Removal of flow divider valve is obvious after an examination of the unit and reference to Fig. IH1069.

NOTE: Valve shown in Fig. IH1074A was used on early series tractors. Valve with lock-out assembly, shown in Fig. IH1074B, replaced the early type valve. Lock-out is used to shut off oil flow to power steering circuit during some stationary operations where increased oil flow (volume) is needed for the auxiliary circuit. On models with lock-out, BE SURE to restore flow to power steering circuit so power steering system will be operative before driving tractor.

To overhaul early type flow divider valve, refer to Fig. IH1074A and proceed as follows: Remove plug (P) and "O" ring, then remove valve spool (V) and spring (S) from body (B). Remove adaptor (A) and "O" ring using a pair of needle-nose pliers, pull safety (relief) valve cartridge (R) and "O" ring from valve body. Inspect spool and bore in body for scoring and/or excessive wear. Scored or worn parts will require renewal of both parts as they are not available separately. Spring (S) should have a free length of 4 5/64 inches and should test 31-36 lbs. at 2½ inches. Do not attempt to disassemble or adjust safety (relief) valve (R). Faulty relief valves are corrected by renewing the unit.

To overhaul the later type flow divider valves, refer to Fig. IH1074B and proceed as follows: Unscrew gland nut and remove lock-out assembly (10) and "O" ring (2), then pull spool (14) and spring (9) from body (8). Remove valve stop (1) and "O" ring (2), then using a pair of needle-nose pliers, pull safety (relief) valve assembly (items 4, 5 and 6) from valve body. Inspect spool and bore in body

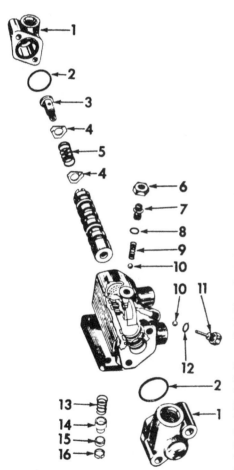

Fig. IH1074—Exploded view of the control (pilot) valve used on series 606 and 2606 International tractors.

1. End cap	8. "O" ring
2. "O" ring	9. Spring
3. Centering spring screw	10. Ball
	11. Plug assembly
4. Centering spring washer	12. "O" ring
5. Centering spring	13. Spring
6. Jam nut	14. Check valve
7. Adjusting screw	15. Seat
	16. Retainer

for scoring and/or excessive wear. Scored or worn parts will require renewal of both parts as they are not available separately. Spring (9) should have a free length of 4 5/64 inches and should test 31-36 lbs. at 2½ inches. Late type units having the poppet type relief valve as shown have parts catalogued separately.

NOTE: Safety valve can be bench tested providing proper equipment such as International Harvester test components FES64-7-6 (pump), FES 64-7-1 (test body), FES64-7-5 (gage), FES64-7-4 (petcock) and FES64-7-2 (plug) is available. Relief valve should open at approximately 1500

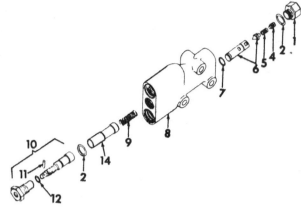

Fig. IH1074B—Exploded view of the late type flow divider valve.

1. Safety valve stop
2. "O" ring
4. Adjusting screw
5. Spring
6. Safety (relief) valve
7. "O" ring
8. Valve body
9. Spring
10. Lock-out assembly
11. Roll pin
12. "O" ring
14. Spool

psi. Bear in mind when testing valves in this manner that pressures will tend to be low due to the small volume of fluid being pumped.

ENGINE AND COMPONENTS

R&R ENGINE WITH CLUTCH

All Models Except 460-606-2606 International

90. To remove the engine and clutch as an assembly, first split (detach) the engine from the clutch housing as outlined in paragraph 189 or 190; then, proceed as follows: Disconnect the radiator hoses. Sling the engine in a hoist and support the steering gear, radiator and side rails assembly. Remove the two engine front mount (timing gear cover) to side rail bolts and lift the engine from side rails.

When reinstalling the engine, first position the engine in the side rails and install one of the cap screws which attach the engine and the side rails to the clutch housing through each of the untapped holes on both sides. Rejoin the engine, side rails and front assembly to the clutch housing and install the attaching cap screws. NOTE: If the tractor is equipped with independent power take-off and/or hydraulic pump, make certain that the drive splines engage clutch cover. With the engine and the side rails attached securely to the clutch housing, detach all hoists, jacks and other methods of support from the tractor. Use a feeler gage to determine the clearance between the engine front mount (timing gear cover) and the side rails. Insert the same thickness of shims as there is clearance between the engine front mount and the side rails; then, install the attaching bolts.

International 460

91. To remove the engine and clutch as an assembly, first split (detach) the engine from the clutch housing as in paragraph 191. Disconnect the radiator hoses, sling the engine in a hoist and support the radiator and front support (bolster) assembly. Remove the cap screws which attach the engine front cover to the front support (bolster) and lift engine away from the front end.

Reinstall the engine in reverse of the removal procedure. If the tractor is equipped with power steering, hydraulic pump and/or independent power take-off, make certain that the drive splines engage the clutch cover correctly as the engine is connected to the clutch housing.

International 606-2606

92. To remove the engine and clutch as an assembly, first split (detach) the engine from the clutch housing as in paragraph 191A. Disconnect the radiator hoses and the radiator brace. If necessary, unclip power steering oil cooler lines at right front, and the power steering cylinder lines at right rear, of engine. Sling engine in a hoist and support the front axle, bolster and radiator assembly, then unbolt bolster (front support) from engine and move engine away from bolster.

Reinstall engine by reversing the removal procedure and bleed power steering system as outlined in paragraph 70.

Fig. IH1075—On non-diesel tractors except 460, 606 and 2606 International it is necessary to block-up the fuel tank bracket at least 2 inches as shown at (A) in order to remove the cylinder head. On diesel tractors, the fuel tank should be blocked-up at least 4 inches at (A).

CYLINDER HEAD

Non-Diesel Models Except 460-606-2606 International

93. To remove the cylinder head, first remove hoods and drain the cooling system. Disconnect ground cable from battery and the fuel line and gage sending wire from fuel tank. Remove the hood center support and fuel tank. Remove the four cap screws which attach fuel tank bracket to the clutch housing; then, block up the bracket at least two inches as shown at (A—Fig. IH1075). Remove the coolant temperature bulb, rocker arm cover, rocker arms assembly and push rods. NOTE: The rocker arms shaft is two-piece. Unbolt the coil bracket from the cylinder head and move the

coil and wiring harness out of the way. Disconnect the coolant by-pass from the thermostat housing. Unbolt the thermostat housing from the cylinder head and move the housing and the fan assembly forward out of the way. Position a block between the right side rail and the cylinder block, unbolt the inlet and exhaust manifolds from the head and allow the manifolds to lower and the carburetor to rest on the block. Remove the remaining cylinder head retaining cap screws and lift off the cylinder head.

The copper side of the cylinder head gasket should face toward top. Guide studs should be used to position the gasket and head. The cylinder head retaining cap screws should be tightened in the sequence shown in Fig. IH1076 and to a torque of 85-95 ft.-lbs. Manifold retaining cap screws should be torqued to 20-30 ft.-lbs.

The cylinder head retaining cap screws should be retorqued after the engine has been run and is at operating temperature. Tappet gap should be adjusted to 0.027 hot for both inlet and exhaust.

Non-Diesel Models 460-606-2606 International

94. To remove the cylinder head on International 460, 606 and 2606 non-diesel tractors, first remove the hoods (front and rear) and drain cooling system. Disconnect ground cable from battery and the fuel line and gage sending wire from fuel tank. Remove the fuel tank. Remove the heat deflector plate, unbolt coil bracket from head and lay on the left side of the engine out of the way. Remove the temperature gage bulb and disconnect the by-pass hose from thermostat housing. Unbolt thermostat housing from cylinder head and move same forward out of the way. Disconnect air cleaner hose, choke cable, throttle rod and fuel line from the carburetor. Remove the intake assembly and outlet pipe from air cleaner. If equipped with underslung exhaust, disconnect exhaust pipe elbow from exhaust manifold. Unbolt and remove the manifolds and carburetor assembly. Unclip wiring loom, and remove the rocker arm cover and the rocker arms assembly. Disconnect wires from spark plugs. Remove the remaining cylinder head retaining cap screws and lift head from tractor.

The copper side of the cylinder head gasket should face toward top. Guide studs should be used to position the gasket and head. The cylinder head retaining cap screws should be tightened in the sequence shown in Fig.

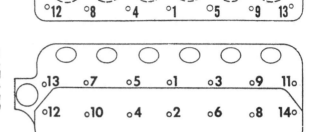

Fig. IH1076 — The non-diesel engine cylinder head retaining cap screws should be tightened in the sequence shown and to a final torque of 85-95 Ft.-Lbs.

Fig. IH1079 — The diesel engine cylinder head retaining cap screws should be tightened in the sequence shown and to a final torque of 110-120 Ft.-Lbs.

IH1076 and to a torque of 85-95 ft.-lbs. Manifold retaining cap screws should be torqued to 20-30 ft.-lbs.

The cylinder head and manifold retaining cap screws should be retorqued to the recommended values after the engine has been run and is at operating temperature. Tappet gap should be adjusted to 0.027 hot for both the inlet and the exhaust.

All Diesel Models

95. To remove the cylinder head on diesel tractors, first remove the hoods (front and rear) and drain the cooling system. If equipped with underslung exhaust, disconnect exhaust pipe elbow from exhaust manifold. Remove the high pressure fuel lines from the injection pump to the nozzles and cap all openings **immediately** to prevent the entrance of dirt. Disconnect and remove the air cleaner inlet and outlet assemblies; then unbolt and remove manifold from left side of head. Disconnect the coolant by-pass hose from thermostat housing. Unbolt thermostat housing from cylinder head and move the housing forward out of the way. Remove the temperature gage bulb from cylinder head and disconnect the glow plug wire from junction block. Remove the tank to filters fuel line and the three cap screws which attach the fuel filters bracket to the head; then, lift off fuel filters assembly.

NOTE: All fuel connections should be capped **immediately** after they are disconnected to prevent the entrance of dirt or other foreign material.

On all diesel models except 460, 606 and 2606 International, proceed as follows: Disconnect the ground cable from the battery and the gage sending wire from the fuel tank. Remove the hood center support and fuel tank. Remove the four cap screws which attach fuel tank bracket to clutch housing; then, block the bracket up at least four inches as shown at (A—Fig. IH1075).

In 460, 606 and 2606 International tractors, proceed as follows: Dis-

connect ground cable from battery and the gage sending wire from fuel tank. Remove the fuel tank and heat deflector plate.

On all diesel models, remove the rocker arm cover and rocker arms assembly. Remove the remaining cylinder head retaining cap screws, attach a hoist and lift off head.

When renewing cylinder head gasket, be sure to use latest type having rubber grommets at front and rear corners on push rod side. When installing cylinder head gasket with the rubber grommets, inspect rocker arm shafts to insure that plugs are included. If not, replace shafts with later type which are plugged as the gasket with the rubber grommets does not provide passage for the trapped oil present around the cap screws at left front and rear corners of cylinder head when the old type rocker arm shafts are used.

Guide studs (second hole from front and rear either side) should be used to position cylinder head during installation. Tighten cylinder head retaining screws in the sequence shown in Fig. IH1079 and to a torque of 110-120 ft.-lbs. Retorque cylinder head retaining screws after engine has been run and is at operating temperature. Tappet gap should be adjusted to 0.027 hot for both inlet and exhaust.

VALVES AND SEATS

Non-Diesel

96. Inlet and exhaust valves are not interchangeable. Inlet valves seat directly in the cylinder head; whereas, the cylinder head is fitted with renewable seat inserts for the exhaust valves which are available in standard size as well as oversizes of 0.015 and 0.030. Valve face and seat angle for both the inlet and the exhaust is 30 degrees. Valve seats can be narrowed if necessary by using a 10 degree stone.

Positive type valve rotators are used on exhaust valves and faulty rotator action is corrected by renewing the rotator unit. Intake valves on C-221 engines prior to serial number 27941, and C-263 engines prior to serial number 25609, were fitted with a seal ring located between valve stem and spring retainer. C-221 engines serial numbers 27941 to 35742 and C-263 engines serial numbers 25609 to 37761 had intake valves only fitted with an umbrella type seal while both engines following these serial numbers had all valves fitted with the umbrella seals.

When removing valve seat inserts, use the proper puller, or pry them out with the edge of a large chisel. Do not attempt to drive chisel under insert or counterbore will be damaged. Chill new seat insert with dry ice or liquid Freon prior to installing and when new insert is properly bottomed, it should be 0.008-0.030 below edge of counterbore. After installation, peen the cylinder head material around the complete outer circumference of the valve seat insert. The O. D. of new standard insert is 1.5655 and insert counterbore I. D. is 1.5625.

Check the valves and seats against the specifications which follow:

Inlet

Head margin, new...............5/64
　renew if less than.............1/32
Stem diameter 0.3715-0.3725
Stem to guide diametral
　clearance 0.003-0.005
Seat width 0.048-0.074
Total valve run-out, less than....0.002
Total valve seat run-out,
　less than 0.002
Tappet gap (hot) 0.027

Exhaust

Head margin, new...............5/64
　renew if less than............1/32
Stem diameter 0.371-0.372
Stem to guide diametral
　clearance 0.0035-0.0055
Seat width 0.083-0.103
Total valve run-out, less than....0.002
Total valve seat run-out,
　less than 0.002
Tappet gap (hot) 0.027

Diesel

97. Inlet and exhaust valves are not interchangeable. Both the inlet and the exhaust valves seat directly in the cylinder head with a face and seat angle of 45 degrees. Valve seats can be narrowed if necessary by using a 15 degree stone. On diesel engines prior to engine serial number 10984, the exhaust valves were equipped with "Roto-Cap" valve rotators; whereas, the inlet valves did not have rotators but were equipped with an umbrella type oil deflector. On diesel engines

serial number 10984 and up, both the inlet and the exhaust valves were equipped with positive type (Rotocoil) valve rotators. Check the valves and seats against the specifications which follow:

Inlet and Exhaust

Stem diameter 0.3715-0.3725
Stem to guide diametral
　clearance 0.0015-0.004
Seat width 0.080-0.090
Total valve run-out, less than...0.002
Total valve seat run-out,
　less than 0.002
Tappet gap (hot)..............0.027

VALVE GUIDES AND SPRINGS

Non-Diesel

98. The inlet and the exhaust valve guides on non-diesel engines are interchangeable. Inlet guides should be pressed in head until top of guide is $1\frac{3}{16}$ inches above the spring recess of the head. The exhaust valve guides should be pressed in head until top of guide is ¾-inch above the spring recess of the head. Guides are pre-sized and if carefully installed should need no final sizing. Inside diameter should be 0.3755-0.3765 and valve stem to guide diametral clearance should be 0.003-0.005 for the inlet, 0.0035-0.0055 for the exhaust.

99. Intake and exhaust valve springs are interchangeable. Springs should have a free length of $2\frac{7}{16}$ inches and should test 146-156 pounds when compressed to a height of $1\frac{19}{32}$ inches. Renew any spring which is rusted, discolored or does not meet the pressure test specifications. Some early production engines had surge dampers in the valve springs, which should be omitted when reassembling.

Diesel

100. The inlet and the exhaust valve guides on diesel engines are interchangeable. Inlet and the exhaust guides should be pressed in head until top of guide is $1\frac{15}{16}$-inch above the spring recess in the head. Guides are pre-sized and if carefully installed should need no final sizing. Inside diameter should be 0.3740-0.3755 and valve stem to guide diametral clearance should be 0.0015-0.004.

101. On early diesel engines which use valve rotators only on the exhaust valves, the inlet and exhaust valve springs are not interchangeable; however, on engines which are equipped with valve rotators on all valves, springs are interchangeable.

The inlet valve spring used on early engines, has 9 coils, a free length of $2\frac{11}{16}$ inches and should exert 149-158 pounds at $1\frac{11}{16}$ inches. The spring used on exhaust valves and the rotator equipped inlet valves, has 7¾ coils, a free length of $2\frac{7}{16}$ inches and should exert 146-156 pounds at $1\frac{19}{32}$ inches. Renew any spring which is rusted, discolored or does not meet the pressure test specifications.

VALVE TAPPETS
(CAM FOLLOWERS)

All Models

102. Tappets are of the barrel type and ride directly in 0.9990-1.0005 diameter bores in the crankcase (cylinder block). Tappet diameter should be 0.9965-0.9970. Oversize tappets are not available. Tappets can be removed from the side of the crankcase after removing valve tappet levers (rocker arms), push rods and side cover plate.

VALVE ROCKER ARM COVER

All Models

103. Removal of the rocker arm cover can be accomplished in the conventional manner on all non-diesel tractors. On diesel tractors removal of the rocker arm cover necessitates removal of the air inlet manifold.

VALVE TAPPET LEVERS
(ROCKER ARMS)

All Models

104. **REMOVE AND REINSTALL.** Removal of the rocker arms assembly may be accomplished in some cases without raising the fuel tank and bracket assembly; however, the rocker shaft bracket hold down screws are also the left row of cylinder head cap screws. Removal of these screws will allow the left side of the cylinder head to raise slightly and damage to the head gasket could result. Because of the possible damage to the head gasket, it is recommended that the cylinder head be removed as in paragraph 93, 94 or 95 when removing the rocker arm assembly.

105. **OVERHAUL.** The rocker arm shaft is two-piece and has an outside diameter of 0.748-0.749. The bushings in the rocker arms are not renewable and the inside diameter of bushings is 0.7505-0.752. The end rocker shaft brackets on all non-diesel engines have a plug and shaft retaining roll pin installed as shown in Fig. IH1081. Diesel engine end rocker shaft brackets have the plugs installed but no shaft retaining roll pin is used. On non-diesel engines, all rocker arms are interchangeable. On diesel en-

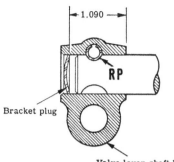

Fig. IH1081 — The roll pin (RP) should be installed with the slot away from the rocker shaft and should engage the smaller notch in the shaft. The bracket plug should be 1.090 from the inner end of the bracket.

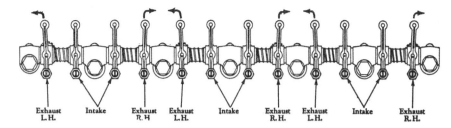

Fig. IH1082—On diesel engines, the inlet rocker arms are interchangeable; but, the exhaust valve rocker arms should be off-set toward the nearest shaft bracket as shown.

gines, all inlet rocker arms are the same; but, the exhaust rocker arms are offset towards the nearest rocker shaft bracket (refer to Fig. IH-1082). On all models, numbers 2, 4 and 6 rocker shaft brackets are provided with dowel sleeves.

VALVE ROTATORS

All Models

106. Positive type valve rotators are installed on the exhaust valves of all non-diesel engines and on diesel engines prior to engine serial number 10984. Valve rotators are used on both the inlet and the exhaust valves of diesel engines, serial number 10984 and up.

Normal servicing of the valve rotators consists of renewing the units. It is important, however, to observe the valve action after engine is started. The valve rotator action can be considered satisfactory if the valve rotates a slight amount each time the valve opens. Valve spring test specifications are listed in paragraph 99 or 101.

VALVE TIMING

Non-Diesel

107. The valves are properly timed when the timing (punched) marks are aligned as shown in Fig. IH1084. The timing marks can be seen after removing the cover as outlined in paragraph 109 or 110.

Diesel

108. The valves and injection pump drive gear are properly timed when the timing (punched) marks are properly aligned as shown in Fig. IH1085. The timing marks can be seen after removing the cover as outlined in paragraph 109 or 110.

TIMING GEAR COVER

Farmall

109. **REMOVE AND REINSTALL.** To remove the timing gear cover on Farmall tractors, first remove steering gear and radiator assembly as outlined in paragraph 14 or 55. Remove the fan, water pump and generator drive belts. Remove the fan blades and generator. Remove cap screws retaining oil pan to timing gear cover and loosen the remaining oil pan screws. Remove the retaining nut and pull the crankshaft pulley from shaft using a suitable puller.

NOTE: On diesel tractors, the crankshaft pulley is also the vibration damper. The damper is cushioned on the pulley hub with rubber; therefore, it is necessary to use a bearing splitter in the front pulley groove and a jaw type puller attached to the bearing splitter to remove the damper and pulleys assembly. If a jaw type puller only is used damage to the damper and pulley assembly may result.

Remove the timing gear cover retaining cap screws, then pull cover forward off dowels and remove from engine. Use care not to damage oil pan gasket.

International

110. **REMOVE AND REINSTALL.** To remove the timing gear cover on International tractors, first remove the grille side sheets and hood skirts and unbolt front of hood from the radiator bracket. Drain coolant and disconnect radiator hoses. Support tractor under clutch housing and sling the radiator and front end assembly in a hoist.

On 460 International tractors, disconnect the drag links from the steering knuckle arms and the radius rod pivot bracket from the clutch housing. Remove the cap screws which attach the front axle support to the timing gear cover and move the complete radiator and axle assembly forward away from the tractor.

On 606 and 2606 International tractors, disconnect power steering oil cooler lines, then separate steering cylinder support, remove cylinder anchor pin and lay steering cylinder aside. Unbolt radius rod pivot bracket, then remove the cap screws which attach front support to timing gear cover and move the complete axle and radiator assembly away from tractor.

On 560 and 660 International tractors, disconnect the drag link from the steering knuckle arm and the radius rod pivot bracket from the clutch housing. Remove the cap screws which attach the front axle support to the side rails and move the complete radiator and axle assembly forward away from the tractor.

On all International tractors, remove the fan, water pump and generator drive belts. Remove the fan blades and generator. Remove cap screws retaining oil pan to timing gear cover and loosen the remaining oil pan screws. Remove the retaining nut and pull the crankshaft pulley from the shaft using a suitable puller.

NOTE: On diesel tractors, the crankshaft pulley is also the vibration damper. The damper is cushioned on the pulley hub with rubber. Therefore, it is necessary to use a bearing splitter in the front pulley groove and a jaw type puller attached to the bearing splitter to remove the damper and pulleys assembly. If a jaw type puller only is used, damage to the damper and the pulley assembly may result.

Remove the timing gear cover retaining screws, then pull cover forward off dowels. Use care not to damage oil pan gasket.

TIMING GEARS

Non-Diesel

111. **CRANKSHAFT GEAR.** The crankshaft gear is keyed and press fitted to the crankshaft. The gear can be removed using a suitable puller after first removing the timing gear cover as outlined in paragraph 109 or 110.

Before installing, heat gear in oil; then, buck-up crankshaft with a heavy bar while drifting heated gear on shaft. Make certain timing marks are aligned as shown in Fig. IH1084.

Fig. IH1084 — The timing marks and gear train on non-diesel tractors. The governor drive gear (G) and the idler gear at (I) are also shown.

Fig. IH1085 — The diesel engine timing gear train and timing marks.

CA. Camshaft gear
CR. Crankshaft gear
ID. Idler gear
IN. Injection pump drive gear

112. **CAMSHAFT GEAR.** The camshaft gear is keyed and press fitted to camshaft. Backlash between the camshaft gear and the crankshaft gear should be 0.004-0.007. The camshaft gear can be removed using a suitable puller after first removing the timing gear cover as outlined in paragraph 109 or 110 and the gear retaining nut.

Before installing, heat gear in oil; then, buck-up camshaft with heavy bar while drifting heated gear on shaft. The gear should butt up against a shoulder on the camshaft. Tighten the gear retaining nut to a torque of 110-120 ft.-lbs. Make certain timing marks are aligned as shown in Fig. IH1084.

113. **IDLER GEAR.** To remove the governor idler gear, it is first necessary to remove timing gear cover as outlined in paragraph 109 or 110. The idler gear shaft is attached to the front of the engine block by a cap screw.

The idler gear shaft diameter should be 2.0610-2.0615 and clearance between shaft and the renewable bushing in gear should be 0.0015-0.0045 with an end play of 0.009-0.013. Make certain that the oil passage in shaft is open and clean.

When reinstalling, make certain that the dowel in the shaft engages the hole in the engine front plate. The shaft retaining cap screw should be torqued to 85-95 ft.-lbs.

Diesel

114. **CRANKSHAFT GEAR.** The crankshaft gear is keyed and press fitted to the crankshaft. The gear can be removed using a suitable puller after first removing the timing gear cover as outlined in paragraph 109 or 110.

Before installing, heat gear in oil; then, buck-up crankshaft with a heavy bar while drifting heated gear on shaft. Make certain timing marks are aligned as shown in Fig. IH1085.

115. **CAMSHAFT GEAR.** The camshaft gear is keyed and press fitted to camshaft. Backlash between the camshaft gear and the crankshaft gear should be 0.003-0.006. The camshaft gear can be removed using a suitable puller after first removing the timing gear cover as outlined in paragraph 109 or 110 and the gear retaining nut.

Before installing, heat gear in oil; then, buck-up camshaft with heavy bar while drifting heated gear on shaft. The gear should butt up against a shoulder on the camshaft. Tighten the gear retaining nut to a torque of 110-120 ft.-lbs. Make certain timing marks are aligned as shown in Fig. IH1085.

116. **IDLER GEAR** To remove the idler gear, it is first necessary to remove timing gear cover as outlined in paragraph 109 or 110. The idler gear shaft is attached to the front of the engine block by a cap screw.

The idler gear shaft diameter should be 2.0610-2.0615 and clearance between shaft and the renewable bushing in gear should be 0.0015-0.0045 with an end play of 0.009-0.013. Make certain that the oil passage in shaft is open and clean.

When reinstalling, make certain that the timing marks are aligned as shown in Fig. IH1085. Make certain that the dowel in the shaft engages the hole in the engine front plate. The shaft retaining cap screw should be torqued to 85-95 ft.-lbs.

117. **PUMP DRIVE GEAR.** To remove the injection pump drive gear it is first necessary to remove the timing gear cover as outlined in paragraph 109 or 110. The drive gear and shaft can be removed by merely withdrawing the unit from injection pump. The gear can be removed from the hub after the three retaining cap screws are removed. The hub can be pressed from the shaft after the retaining nut is removed.

CAUTION: Take care not to damage the double opposed lip seals on the pump drive shaft, the thrust plunger and spring assembly in the forward end of the shaft and/or the bushing in the injection pump housing.

Before installing the pump drive shaft and hub unit, observe rear end of shaft where a small off-center drilled hole is located. During assembly this hole must register with a similar hole in the injection pump rotor or the pump will be 180 degrees out of phase. Lubricate seals with grease and use shim stock to protect the seal lips as shaft is pushed rearward into the injection pump. When installing, make certain that the timing marks on all gears are exactly aligned as shown in Fig. IH1085. Refer to paragraph 146 and retime the injection pump.

CAMSHAFT AND BEARINGS

All Models

118. **CAMSHAFT.** To remove the camshaft, first remove the rocker arms assembly as outlined in paragraph 104 and the timing gear cover as in paragraph 109 or 110. Remove the engine side cover and raise or remove the cam followers (tappets). Working through openings in camshaft gear, remove the camshaft thrust plate retaining cap screws and withdraw camshaft from engine.

Recommended camshaft end play of 0.002-0.010 is controlled by the thrust plate.

Check the camshaft against the values which follow:
Journal Diameter
No. 1 (front)............2.109-2.110
No. 2 ....2.089-2.090
No. 3 ....2.069-2.070
No. 4 (rear)...........1.4995-1.5005

When installing the camshaft, reverse the removal procedure and make certain that the valve timing marks are in register as shown in Fig. IH1084 or IH1085. The camshaft thrust plate retaining cap screws should be torqued to 35-40 ft.-lbs.

119. **CAMSHAFT BEARINGS.** To remove the camshaft bearings, first remove the engine as in paragraph 90, 91 or 92; then, remove camshaft as in the preceding paragraph 118. Remove clutch, flywheel and the engine rear end plate. Extract the plug from behind the camshaft rear bearing and remove the bearings.

NOTE: Oversize dowels are available for engine rear plate on non-diesel engines in cases where dowel holes are enlarged.

Using a closely piloted arbor, install the bearings so that oil holes in bearings are in register with oil holes in crankcase. The chamfered end of the bearings should be installed towards the rear.

The camshaft bearings are pre-sized and if carefully installed should need no final sizing. The camshaft bearing journals should have a diametral clearance in the bearings of 0.0005-0.005.

When installing the soft plug at rear camshaft bearing, use Permatex or equivalent around outer edge to obtain a better seal.

ROD AND PISTON UNITS

All Models

120. Connecting rod and piston assemblies can be removed from above after removing the cylinder head as outlined in paragraph 93, 94 or 95 and the oil pan.

Cylinder numbers are stamped on the connecting rod and cap. Numbers on rod and cap should be in register and face toward the camshaft side of engine. The arrow stamped on the tops of the pistons should point toward front of tractor. Tighten the connecting rod nuts to a torque of 45-55 ft.-lbs.

PISTONS, LINERS AND RINGS

Non-Diesel

121. The cam ground, aluminum pistons are fitted with two compression rings and one oil control ring and are available in standard size only.

When assembling pistons to rods, refer to paragraph 120. Compression rings must be installed with counter bore toward top of piston.

With the piston and connecting rod assembly removed from block, use a suitable puller to remove the dry type cylinder liner (sleeve). Clean the mating surfaces of block and liner before new liner is installed. Top of liner should extend 0.000-0.006 above the top of the cylinder block. Excessive standout will cause leakage at head gasket. Check pistons, rings and sleeves against the values which follow:

Ring end gap
  Compression ....0.010 -0.020
  Oil (Production) .....0.010 -0.018
  Oil (Service),
    rail ....0.015 -0.055
    spacer ....0.018 -0.028
Ring side clearance
  Top compression 0.0035-0.0050
  Second compression ...0.0020-0.0035
  Oil (Production) 0.0025-0.0040
  Oil (Service) 0.0031-0.0074
Cylinder liner I.D., new..3.5603-3.5628
  Renew liner if I.D. exceeds...3.5708
  Renew liner if taper exceeds..0.008
  Liner stand-out 0.000-0.006

Desired diametral clearance between piston (at right angles to piston pin at bottom of skirt) and cylinder liner 0.002-0.003

Diesel

122. The cam ground, aluminum pistons are fitted with two compression rings and two oil control rings and are available in standard size only.

Install the compression rings with the tapered or counterbored side towards the top of the piston. Install the oil rings with the scraper edges towards the bottom of the piston.

With the piston and connecting rod unit removed from block, use a suitable puller to remove the dry type cylinder sleeve (liner). Clean the mating surfaces of block and liner before new liner is installed. Top of liner should be 0.003-0.007 above the top of the cylinder block. Shims of 0.003 and 0.005 thickness are available to control sleeve stand-out. All sleeves should be within 0.002 of having the same stand-out. If more than one shim is used on any one sleeve be sure ends of shims are staggered. Use International Harvester tool FES49, or equivalent to hold sleeves in position when checking the sleeve stand-out. Excessive stand-out will cause leakage at head gasket. Check pistons, rings and sleeves against the values which follow:

Ring end gap
  Top compression ......0.015 -0.025
  Second compression ...0.010 -0.020
  Oil ....0.010 -0.023
Ring side clearance
  Second compression ...0.0025-0.0040
  Oil 0.0025-0.0040
Cylinder liner I.D., new..3.6883-3.6891
  Renew liner if I.D. exceeds...3.6971
  Renew liner if taper exceeds..0.008
  Liner stand-out 0.003-0.007
Desired diametral clearance between piston (at right angles to piston pin at bottom of skirt) and cylinder liner 0.004-0.0056

PISTON PINS

All Models

123. The full floating type piston pins are available in standard and 0.005 oversize. Check piston pin against the values which follow:
Piston pin diameter:
  Non-diesel 0.8748-0.8749
  Diesel (early models).0.8748-0.8749
    (late models) 1.1247-1.1249
Piston pin diametral clearance in piston:
  Non-diesel 0.0002-0.0004
  Diesel 0.0000-0.0003

Piston pin diametral clearance in rod bushing:
  All models 0.0002-0.0005

CONNECTING RODS AND BEARINGS

All Models

124. Connecting rod bearings are of the slip-in, precision type, renewable from below after removing oil pan and rod bearing caps. When installing new bearing shells, make certain that the rod and bearing cap numbers are in register and face toward camshaft side of engine. Bearing inserts are available in standard size as well as undersizes of 0.002, 0.010, 0.020 and 0.030. Check the crankshaft crankpins and the connecting rod bearings against the values which follow:

Crankpin diameter 2.373-2.374
Max. allowable out of round....0.0015
Max. taper 0.0015
Diametral clearance 0.0009-0.0034
Side play 0.007-0.013
Rod bolt torque, ft.-lbs.........45-55

CRANKSHAFT AND MAIN BEARINGS

All Models

125. The crankshaft is supported in four main bearings and the end thrust is taken by the third (rear intermediate) bearing. Main bearings are of the shimless, non-adjustable, slip-in precision type, renewable from below after removing the oil pan and main bearing caps. Removal of the rear main bearing cap on all models, requires removal of the flywheel cover dust plate. Renewal of crankshaft requires R&R of engine. Check crankshaft and main bearings against the values which follow:

Crankpin diameter 2.373-2.374
Main journal diameter.....2.748-2.749
Max. allowable out of round....0.0015
Max. taper 0.0015
Crankshaft end play......0.005-0.013
Main bearing
  diametral clearance ...0.0012-0.0042
  bolt torque, ft.-lbs. 75-85

Main bearings are available in standard size as well as undersizes of 0.002, 0.010, 0.020 and 0.030. Alignment dowels (IH part ED-4000 or equivalent) should be used when installing the rear bearing cap.

NOTE: On engines C-221 at serial number 35094; C-263 at serial number 36694; D-236 at serial number 13754 and D-282 at serial number 44822, the crankshaft thrust bearings (rear intermediate) were changed in that the bearing thrust surface was increased from 3 3/4 to 4 inch diameter. Crankshafts, crankcases and bearing caps were also modified to accept the new bearing and new parts cannot be used in earlier engines. When renewing crankshafts or bearings be sure proper parts are used.

CRANKSHAFT SEALS

Farmall

126. **FRONT.** To renew the crankshaft front oil seal on Farmall tractors, first remove the steering gear and radiator assembly as outlined in paragraph 14 or 55. Remove the generator drive belt. Remove the retaining nut and pull the crankshaft pulley from the shaft using a suitable puller.

NOTE: On diesel tractors, the crankshaft pulley is also the vibration damper. The damper is cushioned on the pulley hub with rubber; therefore, it is necessary to use a bearing splitter in the front pulley groove and a jaw type puller attached to the bearing splitter to remove the damper and pulley assembly. If a jaw type puller only is used, damage to the damper and pulley assembly may result.

Remove and renew seal in the conventional manner. Check the condition of the crankshaft pulley sealing surface and renew or recondition pulley if the surface is not perfectly smooth.

International

127. **FRONT.** To renew the crankshaft front oil seal on International tractors, first remove the front side panels and hood skirts and unbolt front of hood from radiator bracket. Drain coolant and disconnect radiator hoses. Support tractor under clutch housing and sling the radiator and front assembly in a hoist.

On 460 International tractors, disconnect drag links from steering knuckle arms and radius rod pivot bracket from clutch housing. Disconnect radiator brace from engine. Remove the cap screws which attach front axle support to timing gear cover and move the complete radiator and axle assembly forward away from the tractor.

On 606 and 2606 International tractors, disconnect the oil cooler lines, power steering cylinder and the radius rod pivot bracket from clutch housing. Disconnect radiator brace from engine. Remove cap screws which attach front axle support to timing gear cover and move complete axle assembly forward away from the tractor.

On 560 and 660 International tractors, disconnect drag link from steering knuckle arm and radius rod pivot bracket from clutch housing. Remove cap screws which attach front axle support to side rails and move the complete radiator and axle assembly forward away from tractor.

On all International tractors, remove the generator drive belt. Remove the retaining nut and pull the crankshaft pulley from the shaft using a suitable puller.

NOTE: On diesel tractors, the crankshaft pulley is also the vibration damper. The damper is cushioned on the pulley hub with rubber; therefore, it is necessary to use a bearing splitter in the front pulley groove and a jaw type puller attached to the bearing splitter to remove the damper and pulley assembly. If a jaw type puller only is used, damage to the damper and the pulley assembly may result.

Remove and renew the seal in the conventional manner. Check the condition of the crankshaft pulley seating surface and renew or recondition pulley if surface is not perfectly smooth.

All Models

128. **REAR.** To renew the crankshaft rear oil seal, first remove the flywheel as outlined in paragraph 129. The lip type seal can be removed after collapsing same. NOTE: Take care not to damage the sealing surface of the crankshaft as the rear seal is collapsed, removed and new seal is installed. The part number stamped on the seal should face towards rear.

FLYWHEEL

All Models

129. The flywheel can be removed after first splitting engine from clutch housing and removing the clutch as outlined in paragraph 188.

To install the flywheel ring gear, it is necessary to first heat same to approximately 500 deg. F.

The flywheel retaining cap screws should be tightened to 55-65 ft.-lbs. of torque.

OIL PUMP

All Models

130. The gear type oil pump, which is gear driven from a pinion on the camshaft, is accessible for removal after removing the engine oil pan. Disassembly and overhaul of the pump is evident after an examination of the unit and reference to Fig. IH1090. Gaskets between pump cover and body can be varied to obtain the recommended 0.0025-0.0055 pumping (body) gear end play. Refer to the following specifications:

Pumping gears recommended
  backlash 0.003-0.006
Pump drive gear
  recommended backlash ..0.000-0.008
Pumping (body) gear
  end play 0.0025-0.0055
Gear teeth to body
  radial clearance 0.0068-0.0108

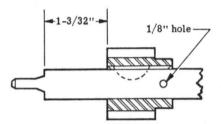

Fig. IH1090 — Exploded view of the engine oil pump. The body gear end play is controlled by the number of gaskets (13). The upper end of the drive shaft is supported by the distributor or the tachometer drive shaft.

2. Drive gear	14. Follower gear
3. Pump drive shaft and body gear	15. Idler shaft
5. Screen	16. Woodruff key
13. Gasket	17. Pin

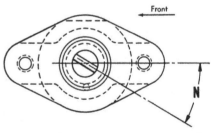

Fig. IH1091—Replacement pump shaft and gear assemblies (3—Fig. IH1090) may need to have a ⅛-inch hole drilled as shown.

Fig. IH1091A—With the number one cylinder on compression stroke and at the static timing position as outlined in text, the oil pump drive gear should be meshed so that angle (N) is approximately 30 degrees.

Drive shaft clearance
  in bore 0.0015-0.003
Pump mounting bolt
  torque, ft.-lbs. 20-23

Service (replacement) pump shaft and gear assemblies are not drilled to accept the pump driving gear pin. A ⅛-inch hole must be drilled through the shaft after the gear is installed on the shaft to the dimensions shown in Fig. IH1091.

Fig. IH1091B—Later diesel engines are fitted with an engine oil cooler which is mounted to left side of engine.

NOTE: When installing the oil pump on non-diesel engines, time the pump as follows: Crank engine until number one piston is coming up on compression stroke and continue cranking until the 2 degree (gasoline) or 3 degree (LP-Gas) BTDC mark on crankshaft pulley is in register with the pointer on the timing gear cover. Install oil pump so that the tang on the pump shaft is in the position shown in Fig. IH1091A. Retime the distributor as outlined in paragraph 180.

OIL PRESSURE RELIEF VALVE

131. On all models, the spring loaded plunger type oil pressure relief valve is located in the oil filter base and is non-adjustable. The spring should be installed with the closed coils in the plunger. Oil pressure should be 30-40 psi at 1800 engine rpm. Check the relief valve plunger and spring against the values which follow:

Valve plunger diameter....0.743-0.745
Plunger diametral clear-
   ance in bore...........0.002-0.007
Spring free length
   (approximately) ....3 inches
Spring test and
   length........18 lbs. @ $1\frac{11}{16}$ inches

ENGINE OIL COOLER
Diesel

131A. All later diesel engines are equipped with an engine oil cooler which is mounted on left side of engine as shown in Fig. IH1091B. Removal is accomplished by disconnecting coolant lines and unbolting manifold from engine (oil filter base).

Removal of relief (by-pass) valve from manifold and manifold from cooler is obvious after an examination of the unit and reference to Fig. IH1091C. Service on assembly consists of renewing faulty parts.

Fig. IH1091C—Exploded view of the diesel engine oil cooler.

  1. Elbow
  2. Connector
  3. Water inlet hose
  4. Water outlet hose
  5. Oil cooler
  6. "O" ring
  7. Gasket
  8. Plug
  9. Gasket
10. Spring
11. Relief (by-pass) valve
12. Manifold

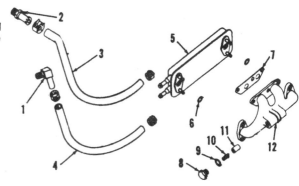

CARBURETOR (GASOLINE)

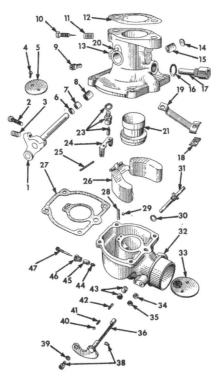

Fig. IH1092—Exploded view of an updraft carburetor typical of that used on non-diesel models.

1. Throttle shaft	27. Gasket
2. Low idle stop screw	28. Idle jet
3. Spring	29. Main air bleed
5. Throttle butterfly	30. Metering nozzle gasket
6. Seal retainer	31. Metering nozzle
7. Dust seal	32. Bowl assembly
8. Bushing	33. Choke butterfly
9. Idle passage plug	34. Dust seal
10. Idle adjusting needle	35. Seal retainer
11. Spring	36. Choke shaft
12. Gasket	38. Swivel
13. Throttle body	40. Ball
14. Plug	41. Friction spring
15. Bushing	42. Stop pin
17. Strainer	43. Drip hole filter
20. Stop pin	44. Main metering seat
21. Venturi	45. Packing
23. Needle valve	46. Packing nut
24. Support	47. Fuel adjusting screw
25. Float pivot	
26. Float	

132. Series 460, 560, 606 and 2606 gasoline tractors are equipped with an IH 1¼ inch updraft carburetor. Series 660 gasoline tractors use an IH 1⅜-inch updraft carburetor. The procedure for disassembling and overhauling the unit is conventional and evident after an examination of Fig. IH-1092. When ordering repair parts be sure to include carburetor number which is located on mounting flange of throttle body.

133. **ADJUSTMENTS.** Before attempting to adjust the carburetor, first start engine and let it run until thoroughly warmed; then, adjust the throttle stop screw (2—Fig. IH1092) to obtain a slow idle speed of 400-450 rpm. Then, turn the idle mixture adjusting screw (10) either way as required to obtain a smooth idle and recheck the slow idle speed. Clockwise rotation of the idle mixture adjusting needle leans the mixture.

Clockwise rotation of the main fuel adjustment screw leans the mixture. The main fuel adjustment screw can be used to reduce the amount of fuel flow when the engine is working under light load conditions; but, when the engine is required to deliver full power, the main adjustment screw must be set five turns open for series 460, 560 and 660; or six turns open for series 606 and 2606.

NOTE: Main adjustment screw seat is calibrated to provide full power mixture and should not be restricted by the main adjustment screw when engine is operating at full power.

Always recheck the idle mixture after making the main fuel adjustment.

134. **FLOAT SETTING.** Distance from farthest face of float to gasket surface of throttle body when inlet needle is closed is $1\frac{5}{16}$-inch. This float setting provides a fuel level of $\frac{9}{16}-\frac{21}{32}$-inch.

LP-GAS SYSTEM

135. On tractors equipped with an LP-Gas fuel system, an Ensign model XG updraft carburetor, with a diaphragm economizer, and a model NS regulator are used. Three adjustments; main (load) adjustment, starting adjustment and throttle stop adjustment are located on the carburetor while the idle mixture adjustment is located on the regulator. The system also incorporates a removable cartridge type fuel filter which can be disassembled and cleaned. See Figs. IH1094A, IH1095A and IH1097.

136. **ADJUSTMENTS.** Before attempting to start engine and adjust system, check and be sure initial fuel adjustments are as follows:

Series 460
Starting adjustment ... 1½ turns open
Main adjustment ..... 2¾ turns open
Idle mixture adjust. ... 1½ turns open

Series 560
Starting adjustment ... 1½ turns open
Main adjustment ..... 3⅜ turns open
Idle mixture adjust. ... 1½ turns open

Series 606-2606
Starting adjustment ... ¾ turns open
Main adjustment ..... 3⅞ turns open
Idle mixture adjust. ... 3 turns open

Series 660
Starting adjustment ... 1½ turns open
Main adjustment ..... 4 turns open
Idle mixture adjust. ... 1½ turns open

With initial adjustments made, start engine and run until engine reaches operating temperature. Place throttle control in low idle position and adjust idle stop screw on carburetor to provide an engine low idle of 425 rpm, then turn the idle mixture needle on regulator either way as required to obtain the highest and smoothest engine operation. Readjust the carburetor idle stop screw, if necessary, to maintain the engine low idle speed of 425 rpm.

With low idle adjustments made, place throttle control in high idle position, turn the main (load) adjustment screw (4—Fig. IH1094A) inward until engine begins to falter, then back-out the main adjustment screw until full power is restored and engine operates smoothly.

NOTE: In some cases, it may be necessary to vary the main adjustment slightly after load is placed on engine.

The initial starting adjustment should provide satisfactory performance; however, it may be varied if cold starting is not satisfactory.

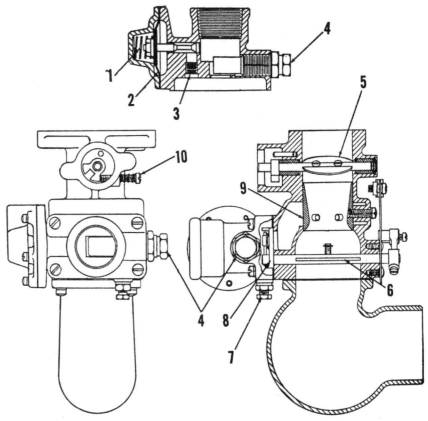

Fig. IH1094—Sectional views of Ensign model XG carburetor used on LP-Gas equipped tractors.

1. Economizer spring	3. Orifice	6. Choke valve	8. Valve
2. Economizer diaphragm	4. Load adjustment	7. Starting adjustment	9. Venturi
	5. Throttle valve		10. Throttle stop screw

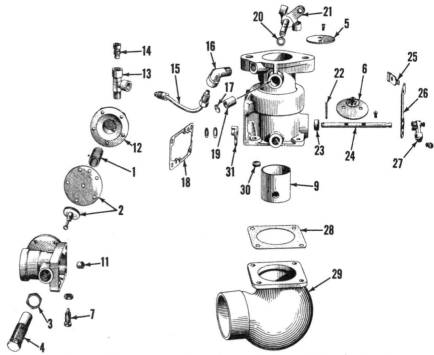

Fig. IH1094A—Exploded view of XG carburetor used on LP-Gas equipped tractors.

1. Economizer spring	7. Starting adjusting screw	16. Elbow	24. Choke shaft
2. Economizer diaphragm	9. Venturi	17. Expansion plug	25. Clamp
3. Lock nut	11. Economizer bleed	18. Gasket	26. Support
4. Load adjusting screw	12. Economizer cover	19. Bushing	27. Choke lever
5. Throttle valve	13. Tee	20. Oil seal	28. Gasket
6. Choke valve	14. Connector	21. Throttle shaft	29. Intake elbow
	15. Tube	22. Pin	30. Expansion plug
		23. Dust washer	31. Valve lever

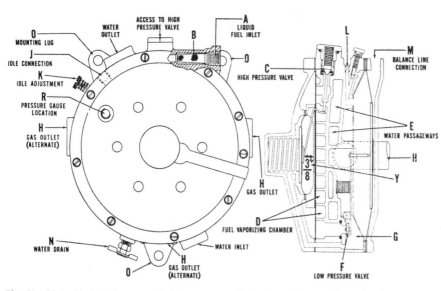

Fig. IH1096 — When setting low pressure valve, center lever with boss "T" when tightening cap screws and set lever to height of boss "T". Valve lever can be bent if necessary to obtain height.

Fig. IH1095—Sectional views of the Ensign model NS regulator used on LP-Gas tractors.

137. CARBURETOR OVERHAUL. The carburetor is serviced the same as a conventional gasoline type carburetor; that is, the carburetor can be completely disassembled, cleared and worn parts can be renewed. Refer to Fig. IH1094A for an exploded view of carburetor. Make certain, however,

that vacuum connections to the economizer chamber do not leak.

138. REGULATOR TROUBLE SHOOTING. If engine will not idle properly, and if turning the idle adjusting screw will not correct the condition, it will be necessary to disas-

semble the regulator unit and thoroughly clean the low pressure valve (F—Fig. IH1095).

To test the condition of the high pressure valve (C), install a suitable pressure gage at the pipe plug connection (R) in the face of the regulator body. If the gage pressure increases after a warm engine is stopped, the high pressure valve is leaking. Under normal operating conditions, the high pressure valve should maintain a pressure of 9-11 psi. If the valve leaks, or does not maintain the proper pressure, renew the high pressure valve.

NOTE: The high pressure valve and its gasket can be serviced without disconnecting or disassembling the regulator unit.

If, after standing for some time, the regulator unit is cold and shows moisture and frost, either the high pressure valve (C) or low pressure valve (F) is leaking or the valve levers are not properly set.

139. REGULATOR OVERHAUL. Disassembly of the regulator is evident after an examination of the unit and reference to Fig. IH1095 and IH1095A. Thoroughly clean all parts and renew any which are excessively worn. When reassembling the unit, make certain that the valve levers are set to the dimensions as follows: The high pressure lever height dimension should be approximately ⅜-inch from top of lever to top of the partition plate as shown in Fig. IH1095.

A boss (T—Fig. IH1096) is machined and marked with an arrow for the purpose of setting the low pressure valve. The valve lever should be centered on the arrow as the lever retaining screws are tightened and the height of the lever should be set to the height of the boss. Lever can be bent if necessary. See Fig. IH1095.

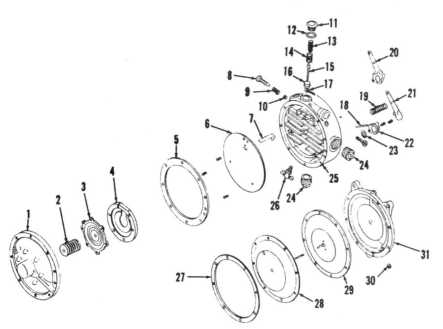

Fig. IH1095A—Exploded view of the model NS regulator used on LP-Gas equipped tractors.

1. Cover	9. Spring	17. Seal	23. Seal
2. High pressure spring	10. Bleed screw	18. Pivot pin	24. Pipe plug
3. High pressure diaphragm	11. Retainer	19. Low pressure valve spring	25. Body
4. Cover	12. Gasket	20. Low pressure valve assembly	26. Drain valve
5. Cover gasket	13. High pressure valve spring	21. Low pressure valve lever	27. Gasket
6. Partition plate	14. Seat retainer spring	22. Low pressure valve seat	28. Partition plate
7. Valve lever	15. High pressure valve		29. Low pressure diaphragm
8. Idle mixture adjusting screw	16. High pressure valve seat		30. Pipe plug
			31. Support plate

140. **LP-GAS FILTER.** Filters used in LP-gas systems should be able to stand high pressure without leakage. Refer to Fig. IH1097. When major engine work is being performed, it is advisable to remove the lower part of the filter, thoroughly clean the interior and renew the treated paper filter cartridge.

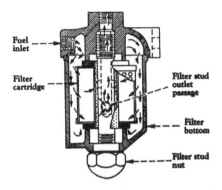

Fig. IH1097—Sectional view of LP-Gas fuel filter. The unit contains a renewable element.

Fig. IH1098—The injection pump should be timed statically using the 3 degrees BTDC mark on series 460, 560, 606 and 2606 tractors. The 5 degree BTDC mark should be used on 660 tractors. Early vibration dampener is shown.

Fig. IH1100 — With the screws (S) loosened, the drive shaft hub can be rotated in relation to the injection pump drive gear due to the elongated holes in the gear.

DIESEL FUEL SYSTEM

Fig. IH1099 — The pump static timing marks can be seen when the timing hole cover is removed.

The diesel fuel system consists of three basic components; the fuel filters, injection pump and injection nozzles. When servicing any unit associated with the fuel system, the maintenance of absolute cleanliness is of utmost importance. Of equal importance is the avoidance of nicks or burrs on any of the working parts.

Probably the most important precaution that service personnel can impart to owners of diesel powered tractors, is to urge them to use an improved fuel that is absolutely clean and free from foreign material. Extra precaution should be taken to make certain that no water enters the fuel storage tanks. This last precaution is based on the fact that all diesel fuels contain some sulphur. When water is mixed with sulphur, sulphuric acid is formed and the acid will quickly erode the closely fitting parts of the injection pump and nozzles.

SYSTEM CHECKS

145. The complete diesel system should be checked as outlined in the following paragraphs whenever the diesel engine does not operate properly.

146. **STATIC TIMING.** To check the injection pump static timing, proceed as follows: Turn the crankshaft until number one piston is coming up on compression stroke and continue turning crankshaft until the pointer extending from timing gear cover is in register with the correct degree mark (3 degrees BTDC for series 460, 560, 606 and 2606 or 5 degrees BTDC for series 660) on the crankshaft pulley. Refer to Fig. IH1098. Shut off fuel, remove timing window cover from side of injection pump and check to be sure that timing marks are aligned as shown in Fig. IH1099.

If pump timing marks are not aligned, loosen the two pump mounting nuts (N—Fig. IH1100A) and turn pump housing either way as required to align the marks; then, retighten the nuts.

In some cases, it may be impossible to turn the injection pump enough to align the timing marks with crankshaft pulley set at the correct degree mark. If this condition is encountered, align the timing pointer with correct degree mark on crankshaft pulley and

remove the gear cover plate from front of timing gear cover. Loosen the three cap screws (S—Fig. IH1100) which attach the pump drive gear to the drive shaft hub. Rotate drive shaft hub in the elongated (adjusting) holes to a point where the pump timing marks (Fig. IH1099) can be perfectly aligned. Retighten the cap screws (S—Fig. IH1100) and reinstall cover plate.

147. **PUMP ADVANCE.** Install the special timing window (TW) as shown in Fig. IH1100A and with No. 1 piston coming up on compression stroke, set the crankshaft at 3 degrees BTDC for series 460, 560, 606 and 2606 or 5 degrees BTDC for series 660. Note (or mark) the position of the pump cam timing line in relation to the marks on the timing window. Start the engine.

With the engine running at low idle no-load speed, the cam timing line should have moved down 2 marks on the window (4 degrees advance) from its static timing position when engine was not running. Advance the throttle to high idle position and with engine running at the recommended high idle no-load speed, the cam timing line should again be 2 marks lower on the timing window (4 degrees advance) than it was with the engine not running. If the cam line isn't 2 marks lower on the window when the engine is running at low and high idle no-load speeds, than it was with the engine stopped, loosen jam nut (JN) and turn the guide stud (GS) as required. Tighten the jam nut and recheck the advance.

Fig. IH1100A—A special transparent timing window (TW) which has timing marks on it should be installed as shown and used to adjust the injection pump timing advance. Early pump is shown, however, later pumps are similar.

1. High speed	2. Low speed	4. Run screw	GS. Governor spring
adjusting screw	adjusting screw	JN. Jam nut	guide stud
	3. Stop screw		N. Nuts

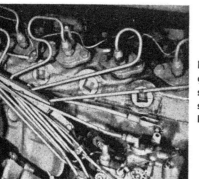

Fig. IH1101 — A vacuum gage can be installed as shown to check for restrictions or air leaks in line between the tank and and the pump.

NOTE: Always check the amount of advance with the jam nut (JN) tight. When the jam nut is loose the guide stud may raise up slightly due to the looseness in the threads and the advance reading may be incorrect.

Accelerate engine by moving the throttle control lever quickly from the low idle speed position to the high idle speed position, and notice the movement of the cam timing line. The cam timing line should move from full **advance** (2 marks or 4 degrees) at low idle speed, to full **retard** (static timing position) during acceleration, then return to full **advance** when the engine reaches high idle speed.

If the cam remains in the advanced position during quick acceleration, back-out the guide stud slightly until

the correct adjustment is obtained.

If a dynamometer is used, the timing line on the cam should be advanced 2 degrees (1 mark from the static position) at 40-60 per cent of rated loaded power with the hand throttle in the high speed position and should be fully retarded (static timing position) at full loaded power.

If guide stud adjustments do not correct the operation of the advance, check for low transfer pump pressure, restriction in the return line, seized advance piston or malfunctioning advance check valve.

148. **VACUUM GAGE TESTS.** To check for a restriction in the line between pump and fuel tank, make certain that fuel tank is full and check the flexible fuel line between pump and second stage filter for being clogged or deteriorated. Install a combination pressure and vacuum gage as shown in Fig. IH1101 and run the engine at high idle rpm. If the gage registers a vacuum more than 10 inches of mercury, check for clogged filters, fuel lines or fittings. Some of the early series 460, 560 and 660 tractors are equipped with a hand primer. When a high vacuum is encountered on these tractors, also check the hand primer for clogging. A vacuum equal to 5-10 inches of mercury indicates that a restriction is beginning.

To check for air leaks between the transfer pump and the fuel tank, a combination pressure and vacuum gage should be installed as shown in Fig. IH1101. Run the engine at low idle speed and close the shut-off valve at tank. When the engine stalls, note the vacuum gage reading, which should be equal to 18-24 inches of mercury.

If the vacuum drops off rapidly, it indicates an air leak. If the vacuum drops off rapidly but an air leak can't be found, check for a stuck pressure regulating valve plunger (78—Fig. IH-1103) or a malfunctioning fuel return check valve (located at the tank end of the fuel return line).

If the vacuum gage reading is less than 18 inches of mercury when the engine stalls but doesn't drop off rapidly; check for a stuck pressure regulating valve plunger (78), damaged "O" rings (76 and 80), weak or broken springs (75 & 79) or damaged transfer pump blades (70).

149. **TRANSFER PUMP PRESSURE.** To check the transfer pump pressure, remove the Allen head plug from bottom of injection pump end plate and install a pressure gage of at least 150 psi capacity in place of

the plug (Fig. IH1102). Start engine, run same at 650 crankshaft rpm and note the pressure reading on the gage, which should be 48-52 psi. With the engine running at 1800 rpm, the gage should register 61-65 psi.

If pressure is low when tested at both 650 and 1800 rpm, remove the plug (81—Fig. IH1103) and install a plug with the next size higher shoulder. Plug (81) is available with shoulder heights from 0.020-0.090 in graduations of 0.010. If varying the shoulder height doesn't remedy incorrect pressure, check for a broken spring (79), a stuck plunger (78) or damaged transfer pump blades (70).

150. FUEL RETURN LINE. To check the fuel return line for being plugged, actuate the hand priming pump on early series 460, 560 and 660 so equipped and listen for fuel return check ball, located at the tank end of the fuel return line, to return to its seat. If seating can be heard, the fuel return is not restricted enough to prevent the engine from starting. If restriction of the fuel return is suspected as the cause of low power, remove and inspect the return line and fittings.

151. NOZZLES. If the engine does not run properly and a faulty nozzle is suspected or if one cylinder is misfiring, locate the faulty nozzle as follows:

Loosen the high pressure line fitting on each nozzle holder in turn, thereby allowing fuel to escape at the union rather than enter the cylinder. As in checking spark plugs in a spark ignition engine, the faulty nozzle is the one which, when its line is loosened, least affects the running of the engine. The malfunctioning nozzle should be removed as outlined in paragraph 156 and tested as in paragraph 155.

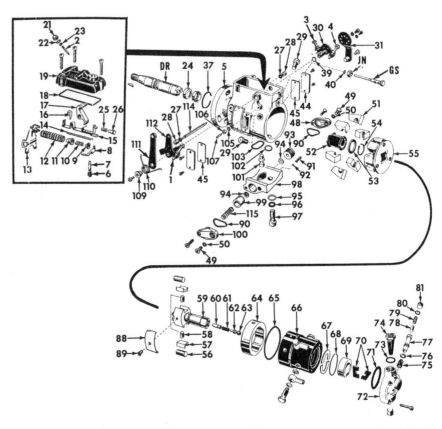

Fig. IH1103—Exploded view of the Roosa Master injection pump used on early series 460, 560 and 660 diesel tractors. Later pumps used on all series are basically similar except that load advance housing (98) is integral with body (5) and torque screw (107) is relocated to aft end of governor control housing. Overhaul should not be attempted by inexperienced personnel.

DR. Drive shaft	13. Governor arm	28. Governor arm	75. Pressure regulating valve
GS. Governor spring guide stud	14. Governor linkage spring	pivot shaft retaining nuts	plunger retaining spring
JN. Jam nut	15. Governor linkage hook	29. Hydraulic head locking screws	76. "O" ring
1. High speed adjusting screw	16. Fuel shut-off cam	30. Fuel shut-off shaft	77. Pressure regulating valve plunger sleeve
2. Low speed adjusting screw	17. Throttle shaft lever	31. Fuel adjusting shut-off arm	78. Pressure regulating valve plunger
3. Shut-off screw	18. Gasket	37. Pilot tube oil seal	79. Pressure regulating valve compression spring
4. Run screw	19. Governor control cover	39. "O" ring	80. "O" ring
5. Housing	21. Nut	40. "O" ring	81. Sleeve retaining plug
6. Spring	22. Washer	44. Timing hole cover	88. Plunger limiting leaf spring
7. Fuel metering valve	23. Seal	45. Gaskets	89. Plunger limiting leaf spring adjusting screw
8. Fuel metering valve arm	24. Drive shaft oil seals	48. Advance end plate (power side)	90. Seals
9. Governor idling spring retainer	25. Governor linkage hook action damper sleeve spring	49. Load advance ball check stop screws	91. Check ball spring
10. Governor idling spring	26. Governor linkage hook action damper sleeve	50. "O" rings	92. Check ball
11. Governor idling spring guide	27. Seals	51. Governor weights (6 used)	93. Load advance power piston
12. Governor control spring		52. Governor thrust sleeve	94. Load advance slide washer
		53. Governor thrust washer	95. Washer
		54. Snap ring	96. Seal
		55. Governor weight retainer	97. Screw
		56. Roller	98. Load advance housing
		57. Shoe	99. Load advance spring piston
		58. Pump plunger	100. Advance end plate (spring side)
		59. Rotor	101. Cam advance screw
		60. Delivery valve	102. Seal
		61. Delivery valve spring	103. "O" ring
		62. Delivery valve stop	105. "O" ring
		63. Delivery valve retaining screw	106. Nut
		64. Cam ring	107. Torque screw
		65. Seal	109. Throttle arm spring retainer
		66. Hydraulic head	110. Throttle arm spring
		67. Distributor rotor retainer	111. Throttle arm
		68. Distributor rotor retainer ring	112. Throttle shaft
		69. Transfer pump liner	114. Governor arm pivot shaft
		70. Transfer pump blades	115. Load advance spring
		71. "O" ring	
		72. Injection pump end plate	
		73. "O" ring	
		74. Fuel strainer	

Fig. IH1102—A pressure gage installed in place of the Allen head plug in the bottom of the injection pump end plate (as shown) can be used to check transfer pump pressure.

Fig. IH1104 — A nozzle test pump and gage similar to that shown can be installed to check for a stuck delivery valve or a scored rotor. Refer to text for testing procedure.

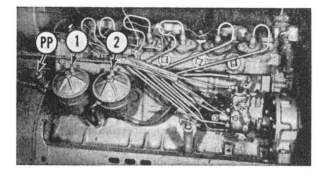

Fig. IH1105—View of the right side of a typical engine. The fuel filter bleed plugs are shown at (1 and 2) and the hand primer pump used on early models at (PP). Late models do not have hand primer.

152. DELIVERY VALVE AND ROTOR. To check the delivery valve and rotor, connect a nozzle test pump and gage to the number one high pressure injection line as shown in Fig. IH1104. Loosen the connection between the nozzle injection line and the test pump hose, actuate the test pump handle until fuel flows out of the loose connection, crank the engine several revolutions with the starting motor until fuel flows out of the loose connection, then retighten the connection. Disconnect the high pressure line from the number 4 or 5 injector. Install the clear timing window (TW) as shown in Fig. IH1100A and turn the engine until the timing line on the governor weight retainer is 2 marks higher on the timing window (4 degrees before the end of injection) than the mark on the cam. This will align the rotor discharge port with the number 1 discharge port in the hydraulic head.

Operate the test pump, maintain a pressure of 2000-2500 psi and record the number of drops of fuel which escape from the disconnected number 4 or 5 injection line in 30 seconds. If more than 25 drops of fuel escape from the disconnected pressure line, the rotor and hydraulic head may be scored or grooved at the discharge ports.

If the amount of escaped fuel is satisfactory for the 30 second interval, operate the test pump until 2500 psi of pressure is built up. If the pressure

drops below 700 psi in 10 seconds, the delivery valve is probably stuck open.

FILTERS AND BLEEDING

The fuel filtering system consists of a fuel strainer and water trap which incorporates the fuel shut-off valve, first stage filter (of the renewable element type) and a second stage filter (of the renewable element type).

153. FILTERS. The water trap and fuel strainer should be removed and serviced regularly and often in order to prevent dirt from contaminating the first and second stage filter elements or entering the injection pump. The primary fuel filter element should have a normal life of about 1000 hours providing the fuel strainer and water trap have been properly serviced. However, in certain areas, the available fuels and operating conditions may decrease filter element life to as little as 250 hours. In these cases, a heavy duty filter element is available which can be used to prolong the element change interval.

The final filter element should last indefinitely if the fuel strainer and water trap and the primary fuel filter have been serviced properly.

A clogged fuel system is usually indicated by the engine misfiring or showing a loss of power. To locate the source of trouble, first clean the fuel strainer and water trap and if trouble persists, renew the primary filter element. If trouble still persists,

renew the final fuel filter element. Fuel system will have to be bled after each of these operations. For information regarding bleeding of the fuel system, refer to paragraph 154.

154. BLEEDING. Each time the filter elements are renewed or if fuel lines are disconnected, it will be necessary to bleed air from the system.

To bleed the fuel filters, open the bleed valve (1—Fig. IH1105) on the top cover of the first stage filter. Open the fuel shut-off valve (incorporated in the fuel strainer and water trap unit). On early models actuate the hand priming pump (PP) until all air has escaped and a solid flow of fuel is escaping from the air bleed hole; then, close the bleed valve. On later models with no hand primer pump, allow fuel to flow by gravity until air free fuel flows from air bleed hole; then close the bleed valve. The second stage filter is also provided with a bleed valve (2) and should be bled in a similar manner.

Normally the injection pump is self bleeding; however, in some cases it may be necessary to proceed as follows:

Loosen the pump inlet line, actuate the hand priming pump on models so equipped and allow fuel to flow from the connection until the stream is free from air bubbles; then, tighten the connection.

Loosen the high pressure fuel line connections at the injectors and crank engine with the starting motor until fuel appears. Tighten the fuel line connections and start engine.

INJECTION NOZZLES

WARNING: Fuel leaves the injection nozzles with sufficient force to penetrate the skin. When testing, keep your person clear of the nozzle spray.

Nozzles of somewhat different construction have been used as shown in Figs. IH1106 and IH1106A. Those shown in Fig. IH1106 were used in engines of series 460, prior to engine serial number D236-9999, as well as in engines of series 560 and 660 prior to engine serial number D282-24415. Those shown in Fig. IH1106A are used in all later series 460, 560 and 660 engines, as well as in all series 606 and 2606 engines.

155. TESTING AND LOCATING FAULTY NOZZLE. If the engine does not run properly, and a faulty injection nozzle is suspected, or if one cylinder is misfiring, locate the faulty nozzle as follows:

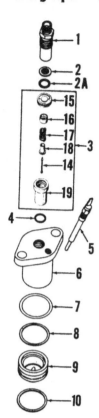

1. Nozzle fitting
2. Screen
2A. Gasket
3. Nozz'e valve
4. Gasket
5. Glow plug
6. Nozzle body
7. Dust seal
8. Gasket
9. Precombustion chamber
10. Gasket
14. Valve (pintle)
15. Nozzle valve cap
16. Retainer
17. Spring
18. Spacer
19. Nozzle valve body

Fig. IH1106—Exploded view of early type injection nozzle.

1. Nozzle fitting
2A. Gasket
2S. Spacer
3. Nozzle valve
4. Gasket
5. Glow plug
6. Nozzle body
7. Dust seal
8. Gasket
9. Precombustion chamber
10. Gasket
11. Nozzle spring seat
12. Spring
13. Valve seat
14. Valve (pintle)

Fig. IH1106A—Exploded view of late type injection nozzle.

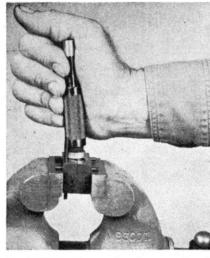

Fig. IH1107—The nozzle valve cap of early injectors can be removed using a special nozzle valve vise and cap removing tool as shown.

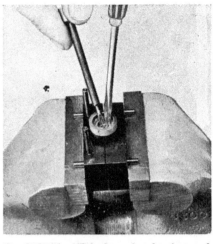

Fig. IH1108—With the valve (early type) mounted in a special vise, two small screw drivers can be used to unlatch the pintle from the retainer.

Loosen the high pressure line fitting on each nozzle holder in turn, thereby allowing fuel to escape at the union rather than enter the cylinder. As in checking spark plugs in a spark ignition engine, the faulty nozzle is the one which, when its line is loosened, least affects the running of the engine.

Remove the suspected nozzle as outlined in paragraph 156, place nozzle in a test stand and check the nozzle against the following specifications:

Early
Opening pressure,
new .... 1525-1625 psi
Opening pressure,
minimum .... 1350 psi
New nozzles should not leak at a pressure less than 1350 psi. The absolute minimum at which a used nozzle should not leak is 1075 psi.

Late
Opening pressure,
new .... 950-1050 psi
Opening pressure,
used (minimum) .... 850 psi
Nozzles should not leak at 700 psi for 10 seconds.

If nozzle requires overhauling, refer to paragraphs 157 or 158.

156. REMOVE AND REINSTALL. To remove any injection nozzle, first remove dirt from nozzle, injection pump and cylinder head; then, disconnect and remove the injector pipe. Cover all openings with tape or composition caps to prevent the entrance of dirt or other foreign materials. Remove the two nozzle retaining stud nuts, lift nozzle from cylinder head and remove the nozzle body dust seal. An OTC HC-689 puller, or equivalent, can be used for withdrawing a stuck nozzle.

When reinstalling, tighten the nozzle hold down nuts evenly and to a torque of 20-25 ft.-lbs.

157. OVERHAUL (EARLY). Remove the nozzle fitting (1—Fig. IH1106) from the nozzle body. Remove the screen (2) and gasket (2A). Remove the nozzle valve unit (3) and gasket (4). Thoroughly clean and inspect all parts, and renew any which are damaged. The screen (2) and all gaskets should be renewed each time the nozzle is subjected to complete or partial overhaul.

The nozzle valve assembly (3) is available as a complete unit. To disassemble and clean the nozzle valve and reset the valve opening pressure, special tools are necessary. Refer to paragraph 157A.

NOTE: It is recommended that the precombustion chambers be removed and cleaned whenever the nozzles are removed for service. Refer to paragraph 159.

157A. To disassemble the nozzle valve assembly for cleaning, refer to Fig. IH1109 and proceed as follows: Clamp the valve assembly in the special nozzle vise (Bacharach tool 66-0116, or equivalent) and pry the nozzle body cap (6) out of the body using Bacharach tool 66-0118, or equivalent (refer to Fig. IH1107). Using a screw driver, press down on the valve retainer until pressure is relieved from upper end of pintle and push upper end of pintle sideways, thereby releasing the valve retainer (refer to Fig. IH1108). Withdraw remaining parts from valve body. Clean all parts in a suitable solvent.

Examine the valve body (1—Fig. IH1109) and cap (6) for scratches, burrs and/or foreign material on the lapped surfaces. The body and cap can be cleaned on a surface plate using mutton tallow.

The pintle (2) and seat can be lapped using Number 400 lapping compound and the special lapping tool (Bacharach tool 66-0115, or equivalent). The instructions for use of the tool should be followed closely.

Spacers (4—Fig. IH1109) with flange thicknesses of 0.020-0.024, 0.026-0.030, 0.032-0.036 and 0.038-0.042 are available to adjust the opening pressures. Opening pressure should be adjusted to 1525-1625 psi and valve should not leak at 225 psi below the opening pressure.

Bacharach tool 66-0121, or equivalent, should be used to press the nozzle cap into position in the body. Refer to Fig. IH1110.

The nozzle fitting should be tightened to 60-70 ft.-lbs. torque.

158. OVERHAUL (LATE). Remove the nozzle fitting (1—Fig. IH1106A) from the nozzle body. Remove gaskets (2A) and spacer (2S). Remove the nozzle valve unit (3) and gasket (4). Thoroughly clean and inspect all parts and renew any which are damaged. The gaskets should be renewed each time the nozzle is subjected to complete or partial overhaul.

NOTE: Early injection nozzles of this type may include a screen instead of gasket (2A). Screen has been replaced with an additional gasket (2A) in later injection nozzles and if a screen is encountered during service, replace it with a gasket.

The nozzle valve assembly (3) is available as a complete unit. To disassemble and clean the nozzle valve and reset the valve opening pressure, refer to paragraph 158A.

NOTE: It is recommended that the precombustion chambers be removed and cleaned whenever the nozzles are removed for service. Refer to paragraph 159.

158A. To disassemble the nozzle valve for cleaning and/or adjusting the opening pressure, press down on the nozzle spring seat (11) until pressure is relieved from upper end of pintle; then, use a screw driver to push upper end of pintle sideways, thereby releasing the nozzle spring seat. Withdraw parts from valve body and clean in a suitable solvent.

The pintle (14) and seat can be lapped using Number 400 lapping compound.

Nozzle spring seats (11) with flange thicknesses of 0.101-0.102, 0.103-0.104, 0.105-0.106, 0.107-0.108, 0.109-0.110,

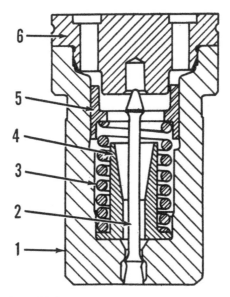

Fig. IH1109—A cross sectional view of the earlier type injection nozzle valve shown in Fig. IH1106.

1. Valve body	4. Spacer
2. Pintle	5. Retainer
3. Spring	6. Valve cap

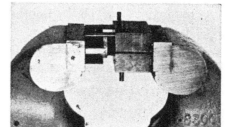

Fig. IH1110—The valve cap of early injector valves should be installed as shown using a special cap installing tool and valve vise.

0.111 - 0.112, 0.113 - 0.114 and 0.115 - 0.116 are available to adjust the opening pressures. Opening pressure should be adjusted to 950-1050 psi and valve should not leak at 700 psi for 10 seconds.

The nozzle fitting should be tightened to 45-50 ft.-lbs. torque.

PRECOMBUSTION CHAMBERS

159. REMOVE AND REINSTALL. Precombustion chambers can be pulled from cylinder head after removing the respective nozzle assembly. The use of a special pre-cup puller may be necessary.

When installing chamber, make certain that side stamped "TOP" is installed toward top of engine.

GLOW PLUGS

160. REMOVE AND REINSTALL. To remove the glow plugs, remove the hood right side sheet (skirt), disconnect the attached wire, unscrew and withdraw plug.

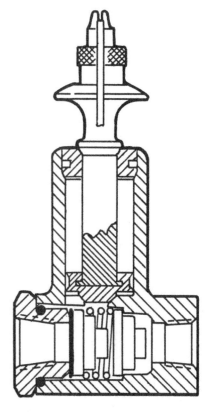

Fig. IH1111—Cross sectional view of typical hand primer pump used on early models. The unit is not used on later models.

PRIMER PUMP

161. CHECK AND OVERHAUL. The pump and check valves on earlier models so equipped can be checked by installing a vacuum gage on the inlet side of the pump, opening a fuel filter bleed valve and actuating the hand primer pump. A pump in good condition will create a vacuum equal to 25 inches of mercury.

Overhaul can be accomplished without removing the pump by removing the inlet line and withdrawing the check valves. Refer to Fig. IH1111. The piston can be withdrawn after first removing the bail.

INJECTION PUMP AND DRIVE

The subsequent paragraphs will outline ONLY the injection pump service work which can be accomplished without disassembly of the injection pump. If additional service work is required, the pump should be turned over to an official diesel service station for overhaul.

162. TIMING. To check the injection pump static timing, refer to paragraph 146. To check the injection pump timing advance refer to paragraph 147.

163. PUMP UNIT—R&R. To remove the complete injection pump unit, first shut off the fuel supply and thor-

Fig. IH1112—View of the injection pump showing the position of the high speed adjusting screw (1) and low speed adjusting screw (2). Adjusting screws (3 and 4) limit the travel of the fuel shut-off arm. Unit shown is an early pump, however, later pumps are adjusted in the same manner.

Fig. IH1113—View of the left (engine) side of the early injection pump showing the torque screw (TS). Late pumps have torque screw located on aft end of governor control housing and advance control housing is integral with main housing.

oughly clean dirt from pump, fuel lines and connections. Turn crankshaft clockwise (viewed from front) until number one piston is coming up on compression stroke and continue turning the crankshaft until pointer extending from timing gear cover is in register with the correct degree mark on the crankshaft pulley (refer to Fig. IH1098). Disconnect the fuel lines and controls, remove the two pump mounting nuts and withdraw pump from engine.

Before reinstalling the pump, remove the timing hole cover from side of injection pump and make certain the pump timing lines are aligned as shown in Fig. IH1099. Mount pump on engine and connect fuel lines and controls.

NOTE: Be very careful when installing the pump over the drive shaft, to prevent damage to the double opposed seals on the shaft.

Recheck the pump timing as in paragraph 146, and bleed the fuel system and in paragraph 154.

164. SPEED ADJUSTMENTS. To adjust the engine governed speeds, first start engine and bring to normal operating temperature. Move the speed control hand lever to the wide open position, loosen the jam nut and turn the high speed adjusting screw (1—Fig. IH1112) either way as required to obtain an engine high idle no-load speed of 1965 rpm for series 460 and 560; 2180 rpm for series 606 and 2606 or 2615 rpm for series 660. Tighten the adjusting screw jam nut. With the high idle speed properly adjusted, the full load engine speed will be 1800 rpm for series 460 and 560 tractors; 2000 rpm for series 606 and 2606 or 2400 rpm for series 660. Move the speed control hand lever to the low idle speed position, loosen the jam nut and turn adjusting screw (2) either way as required to obtain an engine slow idle speed of 625-675 rpm. Tighten the adjusting screw jam nut.

NOTE: It may be necessary to vary the length of the speed control rod in order to obtain full travel in either the low idle or the high idle speed position.

Screws (3 and 4) are provided to set the limits of shut-off arm travel and should not normally require adjustment in the field. Screw (4) adjusts for maximum travel toward the "shut-

off" position; whereas, screw (3) adjusts for "run" position. Adjustment of either screw requires removal of control cover and should be done only by experienced diesel service personnel.

165. TORQUE SCREW ADJUSTMENT. No attempt should be made to adjust the torque screw unless it is determined that engine and its induction system are in good condition and all checks outlined in paragraphs 146 to 152 are made. Also, a dynamometer and an accurate tachometer should be used when making this adjustment.

Load the engine to bring the speed down to rated load speed and hold this load and speed until the engine and fuel temperatures stabilize. Observe the exhaust smoke. Turn the torque screw (TS—Fig. IH1113) as required to obtain a very light colored exhaust smoke. If the air entering the air inlet is at a cold temperature, the adjustment should be made as outlined above, then, the screw should be turned in to reduce the observed horsepower 2 to 5 per cent (depending upon the degree of cold).

NOTE: If fuel delivery cannot be increased (to increase the exhaust smoke) by turning the torque screw, the high idle engine speed, governor linkage hook adjustment, fuel shut-off adjustment and/or the roller to roller dimension (leaf spring adjustment) is probably incorrect.

166. PUMP DRIVE GEAR AND SHAFT—R&R. To remove the injection pump drive shaft, it is first necessary to remove the pump as outlined in paragraph 163. Remove gear front cover plate from the timing gear cover and withdraw the thrust plunger and spring from the pump drive shaft. Mark the relative position of the gear on the pump drive shaft hub. Remove the pump adapter plate and the three screws which attach the gear to the injection pump drive shaft hub. The shaft and hub can be withdrawn from the rear; however, to remove the drive gear it is necessary to remove the timing gear cover as outlined in paragraph 109 or 110. Refer to Fig. IH1114 for timing marks if gear is removed.

When reinstalling the drive shaft, make certain the previously scribed marks on gear and hub are aligned. Refer to paragraphs 162 and 163 for reinstalling and timing the injection pump and paragraph 154 for bleeding the fuel system.

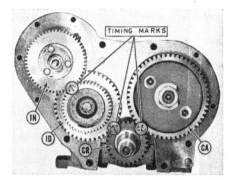

Fig. IH1114—The diesel engine timing gear train and timing marks.

CA. Camshaft gear
CR. Crankshaft gear
ID. Idler gear
IN. Injection pump drive gear

NON-DIESEL GOVERNOR

The governor used on non-diesel tractors is a centrifugal flyweight type and is driven from the crankshaft gear via an idler gear. Two basically similar governors have been used as shown in Figs. IH1115 and IH1116. Note that the primary difference in the two governors is in the control linkage and brackets. On C221 engines, prior to engine serial number 19945, and C263 engines, prior to engine serial number 19081, the bumper spring (28) was not included in the governor.

Before attempting any governor adjustments, check operating linkage and eliminate any binding or lost motion.

ADJUSTMENTS

All Models

170. **SYNCHRONIZING GOVERNOR TO CARBURETOR.** In some cases, servicing or disassembly of carburetor, governor, or operating linkage will change the center to center distance between carburetor throttle shaft lever and governor rockshaft lever. When this occurs, adjust the carburetor control rod as follows to synchronize carburetor and governor. Loosen generator adjusting and mounting bolts, remove belt from pulley and pull generator away from engine. Place the hand throttle lever in high idle position and disconnect carburetor control rod clevis governor rockshaft lever. Hold governor rockshaft lever in the wide open position and adjust control rod clevis until clevis pin can be freely inserted in clevis and rockshaft lever hole, then remove clevis pin and unscrew (lengthen rod) one complete turn. Reconnect clevis to rockshaft lever and tighten jam nut. Reinstall and adjust generator belt. Adjust the engine high idle rpm as outlined in paragraph 171.

171. **HIGH IDLE SPEED.** Synchronize carburetor and governor as outlined in paragraph 170, if necessary. On models so equipped, back out the bumper screw, which is located at aft end of governor, to remove all tension from the bumper spring. Start engine and run until operating temperature is reached. Place hand throttle in high idle position and proceed as follows:

On series 460, 560 and 660 equipped with early type governor (Fig. IH1115), loosen jam nut and turn governor spring adjuster (5) either way as required to obtain the specified high idle rpm which follows:

On series 460, 560 and 660 equipped with late type governor (Fig. IH1116), and the series 606 and 2606, loosen jam nut and turn high idle adjusting screw, located in bracket (4), as required to obtain the following specified high idle rpm.

Series 460 1980
Series 560 1980
Series 606-2606 2200
Series 660 2640

On models equipped with bumper screw, proceed as follows: With engine high idle rpm adjusted, place hand throttle in low idle position, then quickly advance lever to the high idle position. If engine surges excessively (more than twice), loosen jam nut and turn bumper screw in only far enough to eliminate engine surging. Tighten jam nut.

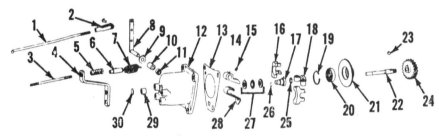

Fig. IH1115—Exploded view of early type governor assembly. Refer also to Fig. IH1116.

1. Carburetor control rod	7. Governor spring	15. Set screw	23. Woodruff key
2. Clevis	8. Rockshaft and lever	16. Governor weight	24. Drive gear
3. Governor control rod	9. Oil seal	17. Thrust sleeve	25. Snap ring
4. Bracket	10. Bearing	18. Weight carrier	26. Snap ring
5. Spring adjuster	11. Bearing	19. Snap ring	27. Thrust bearing
6. Spring retainer	12. Housing	20. Bearing	28. Bumper spring
	13. Gasket	21. Bearing retainer	29. Bearing
	14. Fork	22. Shaft	30. Expansion plug

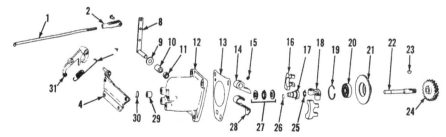

Fig. IH1116—Exploded view of the late type governor assembly. Refer also to Fig. IH1115.

1. Carburetor control rod	11. Bearing	18. Weight carrier	25. Snap ring
2. Clevis	12. Housing	19. Snap ring	26. Snap ring
4. Bracket	13. Gasket	20. Bearing	27. Thrust bearing
7. Governor spring	14. Fork	21. Bearing retainer	28. Bumper spring
8. Rockshaft and lever	15. Set screw	22. Shaft	29. Bearing
9. Oil seal	16. Governor weight	23. Woodruff key	30. Expansion plug
10. Bearing	17. Thrust sleeve	24. Drive gear	31. Speed control lever

NOTE: In some cases, it may be necessary to vary length of governor control rod in order to obtain engine high idle speeds. This can be done on series 460, 560 and 660 by adjusting turnbuckle located just aft of bracket on governor housing. On series 606 and 2606, disconnect aft end of rod from bellcrank and adjust ball joint.

172. **LOW IDLE SPEED.** Run engine until normal operating temperature is reached; then, with engine running, move the governor control hand lever to the low speed position. Turn throttle stop screw on carburetor until the specified low idle speed of 400-450 engine rpm is obtained.

If the engine low idle speed cannot be obtained or maintained, check the governor to carburetor synchronization as outlined in paragraph 170. It may also be necessary to vary the length of the governor control rod to obtain full governor action.

OVERHAUL

173. To remove the governor unit, disconnect carburetor control rod from governor rockshaft lever. On series 460, 560 and 660, separate the governor control rod at turnbuckle. On series 606 and 2606 disconnect governor control rod from speed control lever. Unbolt the governor assembly from the engine front plate.

Disassembly of the removed gover-

nor unit will be self-evident after an examination of the unit and reference to Figs. IH1115 and IH1116.

The governor weights should have a diametral clearance of 0.001-0.010 on the pins. The bearings (20 and 27) should be free from any roughness. The rockshaft lever shaft (8) and the governor drive shaft (22) should rotate freely in bushings (29, 10 and 11). Make certain that the seal (9) is in good condition. The governor lubricating oil passage in the engine front cover and governor housing should be open and clean.

After reassembly and reinstallation is completed, check and adjust the engine speeds as in paragraphs 171 and 172.

COOLING SYSTEM

RADIATOR

175. To remove the radiator, first drain coolant. Remove tractor front hood and radiator bracket side sheets. Disconnect radiator hoses, then unbolt fan shroud. On all models except International 460, 606 and 2606 remove the four cap screws which retain radiator in the bracket and carefully withdraw radiator from bracket.

On International 460, 606 and 2606 disconnect upper radiator braces, then unbolt and remove radiator from front support.

FAN

176. On all models except 460, 606 and 2606 International, the fan is

mounted on a fan drive shaft which is attached to the thermostat housing. Two belts are used, one to drive the water pump and generator and the other to drive the fan.

On 460, 606 and 2606 International tractors, the fan is attached to the water pump drive hub and only one belt is used to drive the water pump, fan and generator.

WATER PUMP

177. Removal procedure will be self-evident and can be accomplished without the removal of the radiator.

Disassembly and overhaul procedures will be evident after an examination of the unit and reference to Fig. IH1120 or IH1121.

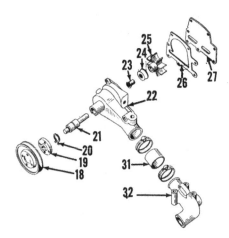

Fig. IH1120—Exploded view of the non-diesel engine water pump.

2. Rear plate	8. Slinger
3. Gasket	9. Shaft and bearing
4. Impeller	10. Snap ring
5. Seal	11. Pulley hub
6. Body	13. Pulley
7. Plug	

Fig. IH1121—Exploded view of the diesel engine water pump.

18. Pulley	25. Impeller
19. Pulley hub	26. Gasket
20. Snap ring	27. Rear plate
21. Shaft and bearing	31. Hose
22. Body	32. Water conector
23. Slinger	and bracket
24. Seal	

IGNITION AND ELECTRICAL SYSTEM

DISTRIBUTOR

180. **INSTALLATION AND TIMING.** With the oil pump properly installed (refer to paragraph 130), crank engine until number one piston is coming up on compression stroke and continue cranking until the 2 degree BTDC (gasoline), or 3 degree BTDC (LP-Gas) mark on crankshaft pulley is in register with pointer on timing gear cover. Install the distributor so that rotor arm is in the number one firing position and adjust the breaker contact gap to 0.020. Loosen the distributor clamp bolt, turn the distributor in the direction opposite to cam rotation until breaker contacts are just beginning to open and tighten the clamp bolt. With the engine running at the correct high idle, no-load speed, the spark should occur 29-31

crankshaft degrees BTDC for 460, 606, 2606 and 560 gasoline models; 24-26 degrees BTDC for 460, 606, 2606 and 560 LP-Gas models; 29-31 crankshaft degrees BTDC for 660 gasoline and LP-Gas models. Firing order is 1-5-3-6-2-4 for all models.

181. **OVERHAUL.** Defects in the battery ignition system may be approximately located by simple tests which can be performed in the field; however, complete ignition system analysis and component unit tests require the use of special testing equipment. Automatic spark advance is obtained by a centrifugal governor built into the unit.

The distributor identification symbol is stamped on the outside diameter of the distributor mounting flange and distributors with symbols Y, AA, AC,

AD, AL and AM have been used. Specifications data and identification information on the distributors used are as follows:

Series 460 and 560 Gasoline Models
Identification symbol ....Y
Breaker contact gap.......0.019-0.023
Breaker arm spring tension, oz...21-25

Advance data in distributor degrees and distributor rpm.
Start advance ....0.0-0.5 @ 200
Intermediate advance ...4.5-6.5 @ 400
Intermediate advance .....9-11 @ 600
Intermediate advance..12.5-14.5 @ 800
Maximum advance ....13.5-14.5 @ 900

Series 460 and 560 LP-Gas Models
Identification symbol ....AA
Breaker contact gap.......0.019-0.023
Breaker arm spring tension, oz...21-25

Advance data in distributor degrees and distributor rpm.
Start advance..............0.1 @ 200
Intermediate advance....3.5-5.5 @ 400
Intermediate advance....7.5-9.5 @ 600
Intermediate advance.....10-12 @ 800
Maximum advance......11-12 @ 900

Series 606 and 2606 Gasoline Models
Identification symbol..............AM
Breaker contact gap...........0.020
Breaker arm spring tension, oz...21-25

Advance data in distributor degrees and distributor rpm.
Start advance...........0-2.0 @ 200
Maximum advance.......16.0 @ 1100

Series 606 and 2606 LP-Gas models
Identification symbol............AL
Breaker contact gap...........0.020
Breaker arm spring tension, oz...21-25

Advance data in distributor degrees and distributor rpm.
Start advance.........0.5-2.5 @ 200
Maximum advance.......12.5 @ 1100

Series 660 Gasoline and LP-Gas Models
Identification symbol.............AC
Breaker contact gap.......0.019-0.023
Breaker arm spring tension, oz...25-29

Advance data in distributor degrees and distributor rpm.
Start advance..........0.0-0.5 @ 200
Intermediate advance....4.5-6.5 @ 400
Intermediate advance.....9-11 @ 600
Intermediate advance..12.5-14.5 @ 800
Maximum advance....13.5-14.5 @ 900

Identification symbol.............AD
Breaker contact gap...........0.020
Breaker arm spring tension, oz...21-25

Advance data in distributor degrees and distributor rpm.
Start advance.............0-1 @ 200
Intermediate advance....3.5-5.5 @ 400
Intermediate advance....7.5-9.5 @ 600
Maximum advance.......11-12 @ 900

GENERATOR AND REGULATOR

182. All models are equipped with Delco-Remy generators and regulators.

Specifications are as follows:

Generator 1100355, 1100356, 1100394, 1100395
Brush spring tension, oz.... 28
Field draw:
  Volts 12.0
  Amperes 1.58-1.67
Cold output:
  Volts 14.0
  Amperes 20.0
  Rpm 2300

Generator 1100391, 1100392
Brush spring tension, oz.... 28
Field draw:
  Volts 12.0
  Amperes 1.50-1.62
Cold output:
  Volts 14.0
  Amperes 25.0
  Rpm 2710

Regulator 1118979, 1118999
Ground polarity .... Negative
Cut-out relay:
  Air gap 0.020
  Point gap 0.020
  Closing voltage, range ... 11.8-14.0
  Adjust to 12.9
Voltage regulator:
  Air gap 0.075
  Setting volts, range 13.6-14.5
  Adjust to 14.0

Regulator 1119241C, D, E
Ground polarity .... Negative
Cut-out relay:
  Air gap 0.020
  Point gap 0.020
  Closing voltage, range ... 11.8-13.5
  Adjust to 12.6
Voltage regulator:
  Air gap 0.062
  Voltage setting 13.9-15.4
  Adjust to 14.6
Current regulator:
  Air gap 0.075
  Current setting, amps. ... 18.5-21.5
  Adjust to 20.0

Regulator 1119270E
Ground polarity .... Negative
Cut-out relay:
  Air gap 0.020
  Point gap 0.020
  Closing voltage, range ... 11.8-13.5
  Adjust to 12.6
Voltage regulator:
  Air gap 0.062
  Voltage setting 13.9-15.4
  Adjust to 14.6

Current regulator:
  Air gap 0.075
  Current settings, amps. .. 23.0-27.0
  Adjust to 25.0

STARTING MOTOR

183. All models are equipped with Delco-Remy starting motors and solenoid switches.

Specifications are as follows:

Starting motor 1107543
Volts 12
Brush spring tension, oz. ... 35
No load test:
  Volts 10.6
  Amperes, min. 75*
  Amperes, max. 100*
  Rpm, min. 6450
  Rpm, max. 8750
Resistance test:
  Volts 5.0
  Amperes, min. 720*
  Amperes, max. 870*
*Includes solenoid.

Starting motor 1107564
Volts 12
Brush spring tension, oz. ... 35
No load test:
  Volts 10.6
  Amperes, min. 105*
  Amperes, max. 200*
  Rpm, min. 6500
  Rpm, max. 14000
Resistance test:
  Volts 3.0
  Amperes, min. 640*
  Amperes, max. 880*
*Includes solenoid.

Starting motor 1107720, 1107744
Volts 12
Brush spring tension, oz. ... 35
No load test:
  Volts 10.6
  Amperes, min. 49*
  Amperes, max. 76*
  Rpm, min. 6200
  Rpm, max. 9400
Resistance test:
  Volts 4.3
  Amperes, min. 270
  Amperes, max. 310
*Includes solenoid

Starting motor 1108642, 1108657
Volts 12
Brush spring tension, oz. 24
No load test:
  Volts 11.8
  Amperes, min. 40
  Amperes, max. 70
  Rpm, min. 6800

Rpm, max. 9200
Lock test:
Volts 5.9
Amperes 615
Torque, ft.-lbs. 29

Starting motor 1113053-1113176
Volts 12
Brush spring tension, oz. ... 48 (1)
No load test:
Volts 11.5
Amperes, min. 57*
Amperes, max. 70*
Rpm, min. 5000
Rpm, max. 7400
Lock test:
Volts 3.4
Amperes 500

Torque, ft.-lbs. 22
*Includes solenoid. (1) 70 oz. on model 1113176

Solenoid switch 1114255, 1114297
Rated voltage 10
Current consumption:
Both windings, volts 10
amperes 47-54
Hold-in windings, volts .. 10
amperes 15.5-17.5

Solenoid switch 1114281, 1114295, 1119923, 1119946
Rated voltage 12
Current consumption:
Both windings, volts 10
amperes 42-49

Hold-in windings, volts .. 10
amperes 10.5-12.5

Solenoid switch 1118821
Rated voltage 12
Current consumption:
Both windings, volts 12
amperes 46-51
Hold-in windings, volts .. 12
amperes .... 1 (max.)

Solenoid switch 1119935
Rated voltage 12
Current consumption:
Both windings, volts 10
amperes 70.4-77.8
Hold-in windings, volts .. 10
amperes 18.20

ENGINE CLUTCH

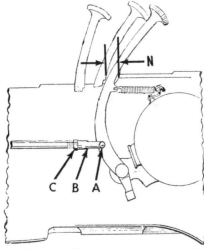

Fig. IH1125—On all models except 460, 606 and 2606 International, the clutch pedal free play should be measured at (N). Refer to text for the specified free play dimension.

Fig. IH1126 — On 460, 606 and 2606 International tractors, the clutch pedal free play should be 1¼-1½ inches when measured at (N).

Fig. IH1126A—To adjust the over center clutch, loosen the adjusting lock and turn the adjusting ring.

ADJUSTMENT

Spring Loaded Clutch (Without Torque Amplifier)

185. Adjustment to compensate for lining wear is accomplished by adjusting the clutch linkage, not by adjusting the position of the clutch release levers.

To adjust the linkage, loosen lock nut (C—Fig. IH1125 or IH1126), remove clevis pin (A) and turn clevis (B) either way as required to obtain the correct free travel (N). The specified pedal free travel is as follows:
Farmall 460 ....1 5/16-1 9/16 inches
Farmall 560 ....1 1/16-1 5/16 inches
Internationals 560 and 660....¾-1 inch
International 460, 606,
2606, ...1¼-1½ inches

Spring Loaded Clutch (With Torque Amplifier)

186. Adjustment to compensate for lining wear is accomplished by adjusting the clutch pedal linkage, not by adjusting the position of the clutch release levers.

The engine clutch linkage and the torque amplifier clutch linkage should be adjusted at the same time. The adjustment procedure is given in paragraph 196, 197, 198 or 198A.

Over Center Clutch

187. When pushing forward on the hand lever to engage the clutch, there should be a definite feel of overcenter action. A pressure (45-55 lbs.) should be felt in the lever as clutch is being engaged, then a definite release of pressure as the clutch goes into engagement.

To adjust the clutch, remove the hand hole cover from bottom of clutch housing and with clutch disengaged, turn engine until lock (Fig. IH1126A) is accessible. Release the lock and turn the adjusting ring clockwise one notch at a time until the specified 45-55 lbs. effort applied to the control lever will engage the clutch. Re-engage the adjusting ring lock. If lock and notch in adjusting ring are not aligned, turn adjusting ring clockwise until alignment is obtained.

187A. On tractors equipped with over center clutch and "Torque-Amplifier", the TA clutch must also be adjusted as outlined in paragraphs 187B and 187C.

NOTE: International 460 Grove tractors, and International 560 and 660 series tractors may be equipped with a single lever clutch control system, and in addition, International 560 and 660 series tractors may also be equipped with a two lever clutch control system. For information pertaining to single lever systems, refer to paragraph 187B. For information pertaining to two lever systems, refer to paragraph 187C.

187B. SINGLE LEVER SYSTEM. Before adjusting the "Torque-Amplifier" clutch, adjust engine clutch as outlined in paragraph 187. Now place 'Torque-Amplifier" control handle in the forward (direct drive) position and loosen cap screw (B—Fig. IH-1126B). Hold down on cap screw (B), and at the same time, hold operating link (under cap screw B) up; then, tighten cap screw (B). This will remove any linkage free travel which may be present. Now loosen cap screw (C), make sure control lever is fully forward, then insert a punch through hole provided in TA clutch adjustment plate, move TA clutch shaft lever rearward until release bearing contacts TA clutch fingers, then move lever ⅛-inch forward and tighten the cap screw (C). This provides ⅛-inch clearance between release bearing and clutch fingers and 2½ inches of free handle travel.

Note: Be sure that release bearing is contacting clutch fingers rather than cap screw (C) butting against end of slot. If cap screw is hitting end of slot, the slot must be lengthened.

Place control lever in the middle (TA drive) position and check the distance between cam plate retainer washer and top end of cam plate slot. This distance should be $\frac{7}{16}$-inch as shown in Fig. IH1126B. If necessary, remove clevis pin (D) and adjust control rod clevis to obtain this measurement.

If deemed necessary, adjustments can be checked by removing the TA housing top cover and measuring between TA clutch disc and pressure plate with control lever in middle (TA drive) position. This clearance should be 0.018 minimum.

187C. TWO LEVER SYSTEM. To adjust the "Torque-Amplifier" clutch on tractors equipped with two levers, first adjust engine clutch as outlined in paragraph 187, then use procedure given in paragraph 197A, to adjust the "Torque-Amplifier" clutch.

REMOVE AND REINSTALL
All Models

188. To remove the engine clutch, it is first necessary to detach (split) engine from clutch housing as outlined in paragraph 189, 190, 191 or 191A. The clutch can then be unbolted and removed from flywheel in the conventional manner.

TRACTOR SPLIT
Farmall 460 and 560

189. To detach (split) engine from clutch housing, first remove hoods,

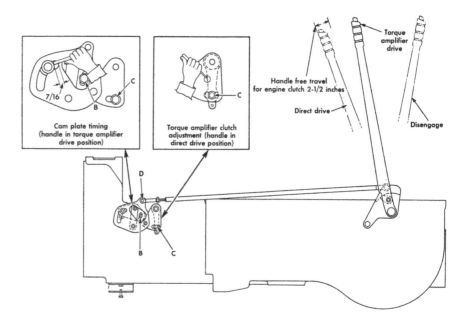

Fig. IH1126B—To adjust the "Torque Amplifier" clutch on models with over-center engine clutch refer to the drawing above and text.

then disconnect the steering shaft center universal joint and unbolt the steering shaft bracket from clutch housing. Drain cooling system and on models so equipped, disconnect the radiator shutter control rod and the tachometer cable. On all models, disconnect the air inlet pipe, heat indicator sending unit, fuel lines, oil pressure gage line, wiring harness and controls from engine and engine accessories. On models equipped with power steering, remove the pressure and return lines that connect to the control valve, hydraulic controls bracket and the pump manifold. Attach hoist to engine half of tractor in a suitable manner and securely block rear half of tractor so it will not tip. Unbolt engine and side rails from clutch housing and separate the tractor halves.

International 560 and 660

190. To detach (split) engine from clutch housing, first remove hoods, drain cooling system and disconnect the head light wires. On models so equipped, disconnect the shutter control rod and head light wires. On all models, disconnect the steering drag link from steering (knuckle) arm and the radius rod pivot bracket from clutch housing. Disconnect the tachometer cable, air inlet pipe, heat indicator sending unit, fuel lines, oil pressure gage line, wiring harness and controls from engine and engine accessories. Attach hoist to engine half of tractor in a suitable manner and securely block rear half of tractor so it will not tip. Unbolt engine and side

rails from clutch housing and separate the tractor halves.

International 460

191. To detach (split) engine from clutch housing, first drain cooling system and remove hoods. Disconnect head light wires, radius rod pivot bracket from clutch housing and both drag links from the steering (knuckle) arms. Remove the fuel tank. Disconnect the heat indicator sending unit, fuel lines, oil pressure line, wiring harness and controls from engine and engine accessories.

Attach hoist to engine half of tractor in a suitable manner and securely block rear half of tractor so it will not tip. Unbolt engine from clutch housing and separate the tractor halves.

NOTE: The two long bolts retaining clutch housing to top of engine should be unscrewed gradually as engine is moved forward. This procedure eliminates the need of removing the steering gear unit and usually saves considerable time.

International 606-2606

191A. To detach (split) engine from clutch housing, first drain cooling system and remove hood skirts, precleaner and hood. Remove air cleaner and disconnect tachometer cable. Disconnect battery cable from starting motor solenoid, wires from generator, resistor, oil pressure switch and headlights. Unclip wiring loom from engine. On non-diesel models, shut off fuel, disconnect choke control, fuel supply line and throttle rod from carburetor. On diesel models, disconnect shut-off control and control rod from

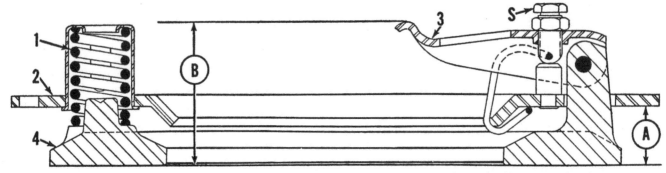

Fig. IH1128 — Sectional view of a spring loaded clutch cover, showing the release lever adjustment dimensions. (A) dimension is the position of the pressure plate in relation to the cover plate which must be maintained when adjusting the release lever height (B).

S. Lever adjusting screw   1. Pressure spring   2. Cover plate   3. Release lever   4. Pressure plate

injection pump, then shut off fuel and disconnect fuel supply line from primary filter. On all models, disconnect steering cylinder lines from control (pilot) valve and lay lines forward. Loosen clip at clutch housing and disconnect aft ends of oil cooler lines. Disconnect temperature indicator bulb from cylinder head. Remove fuel tank sediment bowl. Remove nuts from studs of fuel tank front support and push fuel tank upward off studs. Unbolt stay rod bracket from clutch housing and remove cap screws from lower half of clutch housing. Loop a chain under rear of engine, install a spreader bar over top of fuel tank, then attach a hoist to chain and take weight of engine. Support rear frame, wedge front axle, then remove the remaining clutch housing cap screws and separate tractor.

NOTE: When removing the two long cap screws at top rear of clutch housing, piping may interfere. If this occurs, unscrew cap screws gradually as the engine moves forward.

OVERHAUL CLUTCH

All Models

192. The procedure for disassembly, adjusting and/or overhauling the removed clutch cover assembly is conventional. Overhaul data are as follows:

OVERHAUL DATA

Dimensions (A) and (B) are shown in Fig. IH1128.

Series 460-606-2606 (International Harvester Clutch)

Cover setting (A)..............0.851"
Lever height (B)....2 13/64-2 15/64"
Springs
   Number used 9
   Lbs. test @ height
      Regular ....140 @ 1.44"
      Heavy duty ....165 @ 1.44"
   Free length
      Regular ....2.75"
      Heavy duty ....2.61"

Series 460-606-2606 (Rockford Clutch)

Cover setting (A)..............0.851"
Lever height (B)....2 13/64-2 15/64"
Springs
   Number used 9
   Lbs. test @ height
      Regular ....143 @ $1\frac{7}{16}$"
   Free length ....2.75"

Series 560 (International Harvester Clutch)

Cover setting (A)..............1.020"
Lever height (B)....2 19/64-2 21/64"
Springs
   Number used 12
   Lbs. test @ height
      Regular ....137 @ $1\frac{13}{16}$"
      Heavy duty ....155 @ $1\frac{13}{16}$"
   Free length
      Regular ....$3\frac{5}{32}$"
      Heavy duty ....$2\frac{13}{16}$"

Series 560 (Rockford Clutch)

Cover setting (A)..............1.020"
Lever height (B)....2 19/64-2 21/64"
Springs
   Number used 12
   Lbs. test @ height
      Regular ....140 @ $1\frac{13}{16}$"
      Heavy duty ....155 @ $1\frac{13}{16}$"
   Free length
      Regular ....$3\frac{5}{32}$"
      Heavy duty ....$2\frac{13}{16}$"

Series 660 (Rockford Clutch)

Cover setting (A)..............1.020"
Lever height (B)....2 19/64-2 21/64"
Springs
   Number used 12
   Lbs. test @ height
      Regular ....140 @ $1\frac{13}{16}$"
      Heavy duty ....155 @ $1\frac{13}{16}$"
   Free length
      Regular ....$3\frac{5}{32}$"
      Heavy duty ....$2\frac{13}{16}$"

NOTE: Current series 606 and 2606 tractors are being equipped with the Dyna-Life clutch disc and the heavy duty clutch is no longer available. Heavy duty clutch pressure plate assemblies can be converted for use with the Dyna-Life disc by replacing the six heavy duty springs with nine regular duty springs.

Fig. IH1129—The bearing cage (26), pto drive shaft (25) and clutch shaft (24) removed. The pto drive shaft rear bearing remains in the housing.

CLUTCH SHAFT

All Models (Without Torque Amplifier)

193. To remove the engine clutch shaft on models without torque amplifier, detach (split) engine from clutch housing as outlined in paragraph 189, 190, 191 or 191A; then, withdraw the clutch shaft forward and out of clutch housing.

All Models (With Torque Amplifier)

194. To remove the clutch shaft on all models with a torque amplifier, first detach (split) engine from clutch housing as outlined in paragraph 189, 190, 191 or 191A and remove the engine clutch release bearing and shaft. On models with independent power takeoff, unbolt bearing cage (26—Fig. IH1129) from clutch housing and withdraw the independent power take-off drive shaft (25), bearing cage (26) and clutch shaft (24). Note: A tapped hole is provided in the end of the clutch shaft to aid in its removal. The pto drive shaft rear bearing remains in the housing. To remove this bearing it is necessary to first remove the pto driven gear.

TORQUE AMPLIFIER UNIT

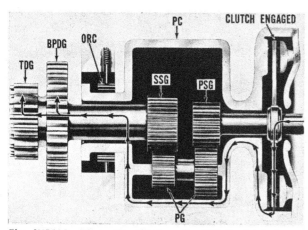

Fig. IH1132—When the "Torque Amplifier" clutch is engaged, the system is in direct drive.

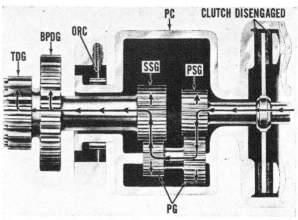

Fig. IH1133—When the "Torque Amplifier" clutch is disengaged, an overall gear reduction of approximately 1.48:1 is obtained.

Torque amplification is provided by a planetary gear reduction unit located between the engine clutch and the transmission. The unit is controlled by a single plate, spring loaded clutch. When the clutch is engaged as in Fig. IH1132, engine power is delivered to both the primary sun gear (PSG) and the planet carrier (PC). This causes the primary sun gear and the planet carrier to rotate as a unit and the system is in direct drive. When the clutch is disengaged as shown in Fig. IH1133, engine power is transmitted through the primary sun gear to the larger portion of the compound planet gears (PG), giving the first gear reduction. The second gear reduction is provided by the smaller portion of the compound planet gears driving the secondary sun gear (SSG). As a result of the two gear reductions, an overall gear reduction of approximately 1.48:1 is obtained.

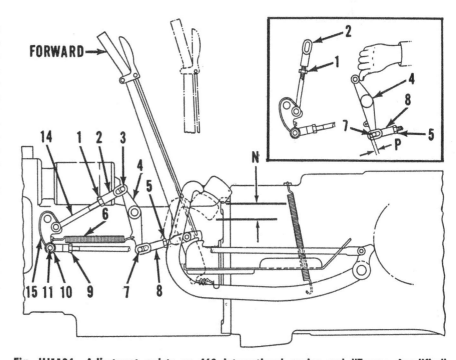

Fig. IH1134—Adjustment points on 460 International engine and "Torque Amplifier" clutch linkage. Series 606 and 2606 are adjusted in the same manner.

T. A. CLUTCH

Models With Spring Loaded Engine Clutch

195. **ADJUST.** Adjustment to compensate for lining wear is accomplished by adjusting the clutch actuating linkage, not by adjusting the position of the clutch release levers. The engine clutch linkage and the torque amplifier clutch linkage should be adjusted at the same time. For 460, 606 and 2606 International tractors, refer to paragraphs 196 and 196A; for 460 and 560 Farmall and 560 International models, refer to paragraphs 197 and 197A; for 660 International models with the foot operated T. A. refer to paragraph 198.

196. 460 - 606 - 2606 INTERNATIONAL. Refer to Fig. IH1134, remove spring (6), loosen lock nut (1) and remove clevis pin (3). Loosen nut (9), remove clevis pin (11) and turn clevis (10) either way as required to obtain a pedal free travel (N) of 1⅜ inches as shown.

196A. After the engine clutch linkage is properly adjusted, place the torque amplifier control lever in the forward position as shown and proceed to adjust the torque amplifier clutch linkage as follows: Loosen lock nut (5), remove pin (7) and turn lever (4) counter-clockwise as far as possible without forcing. See inset. This places the T. A. clutch release bearing against the clutch release levers. Now, adjust clevis (8) to provide a space (P) of $\frac{3}{16}$-inch between the inserted pin (7) and the forward end of the elongated hole in clevis (8). Tighten lock nut (5) and reinstall spring (6). Adjust the length of rod (14) with clevis (2) so that rod (14) is the shortest possible length that will not change the position of levers (15 and 4) when pin (3) in inserted.

197. 460 - 560 FARMALL AND 560-660 (hand operated) INTERNATIONAL. Refer to Fig. IH1135. Remove spring (6), loosen locknut (1) and remove clevis pin (3). Loosen lock nut (9) and remove clevis pin (11). Turn clevis (10) until clutch pedal free travel (N) is as follows:

Farmall 460 .... $1\frac{5}{16}$-$1\frac{9}{16}$ inches

Farmall 560 .... $1\frac{1}{16}$-$1\frac{5}{16}$ inches

International 560 and 600 ... $\frac{3}{4}$-1 inch

Dimension (N) is measured horizontally from point of contact of clutch pedal lever and rear frame cover. After adjustment is complete, tighten lock nut (9).

After the engine clutch pedal linkage is properly adjusted, place the torque amplifier control lever in the forward position as shown and proceed to adjust the torque amplifier clutch linkage as follows:

197A. Loosen lock nut (5), remove clevis pin (7) and turn lever (4) counter-clockwise as far as possible without forcing. See inset. This places the TA clutch release bearing against the clutch release levers. Now, adjust clevis (8) to provide a space (P) of $\frac{3}{16}$-inch between the inserted pin (7) and the forward end of the elongated hole in clevis (8). Tighten lock nut (5) and reinstall spring (6). Adjust the length of rod (14) with clevis (2) so that rod (14) is the shortest possible length that will not change the position of levers (4 and 15) when pin (3) is inserted.

198. INTERNATIONAL 660 (with foot operated Torque Amplifier). On 660 International models with foot-operated torque amplifier, a sheet metal gage $\frac{7}{8}$ - inch wide and $3\frac{1}{8}$ inches long should be used to adjust the engine and TA clutches as follows: Refer to Fig. IH1136 and loosen the cap screws (1 and 2). Depress the clutch pedal and block its return with the $3\frac{1}{8}$ inch length of the gage as shown in Inset "B". Move lever (A) forward until the engine clutch release bearing just contacts the clutch release levers, then retighten cap screw (1) and remove the $3\frac{1}{8}$ inch gage.

Depress the pedal and block its return with the $\frac{7}{8}$-inch width of the gage as shown in Inset "C". Move the lever (B) to the rear until the TA clutch release just contacts the clutch release levers, then retighten cap screw (2) and remove the $\frac{7}{8}$-inch gage.

If the unit has been disassembled, place a straight edge (C) on top of the cam roller and along both punch marks on the cam plate. If the top edge of the roller and the two punch marks aren't aligned, vary the length

Fig. IH1135—Adjusting points for the engine and "Torque Amplifier" clutch linkage. This illustration is typical of all models except 460, 606 and 2606 International tractors, tractors with over center engine clutch and 660 models with foot operated TA clutch.

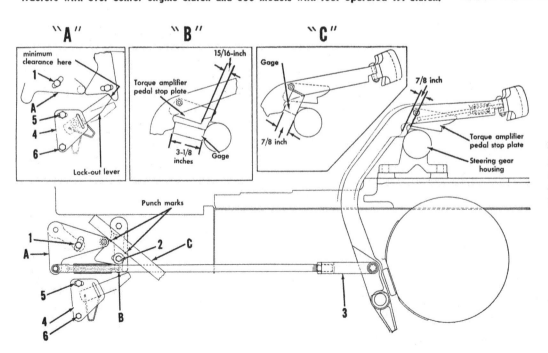

Fig. IH1136 — Adjustment points for the engine and "Torque Amplifier" clutch linkage on 660 tractors with foot operated TA clutch.

of the control rod at clevis (3) and repeat the clutch adjustments until the marks align.

To adjust the TA lock-out plate (4—Inset "A"), loosen the cap screws (5 and 6) and hold the clutch pedal in the completely depressed position. Adjust the plate (4) until the lock-out lever can move in and out of position on the cam plate freely yet with a minimum of clearance. Recheck the clutch adjustments.

198A. On 660 International models with hand-operated torque amplifier, refer to paragraph 197 and 197A.

Models With Over Center Engine Clutch

199. **ADJUST.** To adjust the over center engine clutch and the TA clutch on models so equipped, refer to paragraphs 187 through 187C.

All Models

200. **R&R AND OVERHAUL.** To remove the torque amplifier clutch cover assembly and lined plate, first detach (split) engine from clutch housing as outlined in paragraph 189, 190, 191 or 191A. and proceed as follows:

On Farmall tractors equipped with a side mounted belt pulley, remove the belt pulley unit. On all models except 460, 606 and 2606 International, remove the starter, fuel tank sediment bowl and clutch housing top cover.

On 460 International tractors, first remove the steering gear and fuel tank assembly as in paragraph 23 or 65. On 460 International tractors so equipped, remove the hydraulic system control valves from the top of the clutch housing. Remove the clutch housing top cover.

On 606 and 2606 International tractors, remove battery and starting motor. Disconnect hydraulic lines from control valve and flow divider valve. Disconnect hydraulic lift pressure line. Unbolt fuel tank support and steering gear support and using a hoist, lift fuel tank, Hydrostatic hand pump assembly and instrument panel assembly from tractor. Remove clutch housing top cover.

On all models, remove the clutch shaft as in paragraph 194. Disconnect linkage from the TA clutch release shaft and remove lock screw from the TA clutch release fork. Remove snap ring from right end of the TA clutch release shaft; then, withdraw shaft, Woodruff keys, fork and release bearing and carrier. Use three $\frac{5}{16}$"-18 by $\frac{7}{8}$" cap screws and plain washers and screw them into the tapped holes provided in the pressure plate to keep the assembly under compression; then, unbolt the TA clutch cover assembly (35—Fig. IH1138) from carrier and withdraw the clutch cover assembly and lined plate.

If carrier (47—Fig. IH1139) is to be removed, bend tang of locking washer (46) out of notch in nut (45), remove spanner nut (Fig. IH1140) and bump carrier from splines of the primary sun gear.

Examine the driven plate for being warped, loose or worn linings, worn hub splines and/or loose hub rivets. Disassemble the clutch cover assembly and examine all parts for being excessively worn. The six pressure springs should have a free length of $2\frac{1}{32}$ inches and should require 156-166 lbs. to compress them to a height of $1\frac{1}{4}$ inches. Renew pressure plate if it is grooved or cracked. Renew back plate (44—Fig. IH1138) if it is worn around the drive lug windows.

When reassembling, adjust the release levers to the following specifications. With a back plate to pressure plate measurement (K—Fig. IH1141) of $\frac{13}{32}$-inch, the release lever height (L) from friction face of pressure plate to release bearing contacting surface of release levers is $1\frac{5}{8}$ inches

When reassembling, observe the clutch carrier and back plate for balance marks which are indicated by an arrow and white paint. If the balance marks are found on both parts, they should be assembled with the marks as close together as possible. If no marks are found, or if only one part is marked, the clutch balance can be disregarded. Install the remaining parts by reversing the removal procedure.

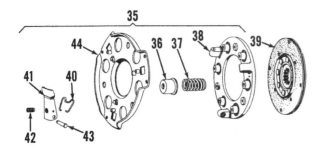

Fig. IH1138 — Exploded view of the "Torque Amplifier" clutch. Later discs (39) have bonded metallic facings and pin (43) is retained by "E" ring.

36. Spring cup
38. Pressure plate
39. Driven disc
40. Lever spring
41. Release lever
42. Adjusting screw
43. Lever pin

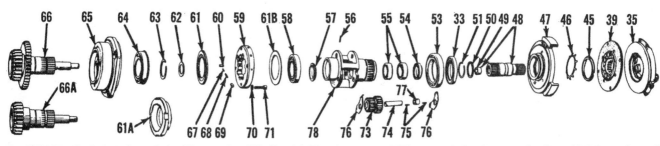

Fig. IH1139—Exploded view of the "Torque Amplifier" unit. Planetary gears (73) are available in sets only. Item (66) is used on all tractors except 460, 606 and 2606 International; item (66A) is used on 460, 606 and 2606 International tractors; item (61) is used on 560 and 660 tractors; item (61A) is used on 460, 606 and 2606 tractors.

33. Oil seal	55. Needle bearings	64. Bearing	67. Pin
35. Clutch cover assembly	56. Roll pins	65. Transmission drive	68. Spring
39. Driven plate	57. Spacer	shaft bearing cage	69. Plug
45. Nut	58. Bearing	66. Transmission drive	70. Retaining ring
46. Locking washer	59. Over-running clutch	shaft, secondary sun	71. Special screw
47. Clutch carrier	ramp	gear and side	73. Planetary gears
48. Primary sun gear	60. Clutch roller	mounted belt pulley	74. Gear shaft
49. Needle bearing	61. Rear thrust washer	drive gear	75. Needle bearing
50. Thrust washer	61A. Rear thrust washer	66A. Transmission drive	76. Thrust plate
51. Snap ring	61B. Front thrust washer	shaft and secondary	77. Bearing spacer
53. Bearing	62. Thrust washer	sun gear	78. Planet carrier
54. Oil seal	63. Snap ring		

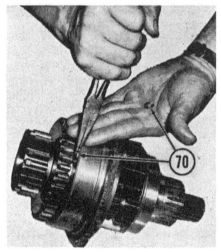

Fig. IH1140 — Installing the spanner nut. The clutch carrier should be blocked to prevent its turning. Remove the nut using a similar procedure.

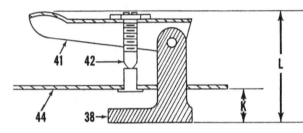

Fig. IH1141 — When adjusting the release lever height (L) of 1⅝ inches, the back plate to pressure plate measurement (K) of 19/32 inch be maintained.

Fig. IH1142 — The retaining rings (70) should be installed as shown.

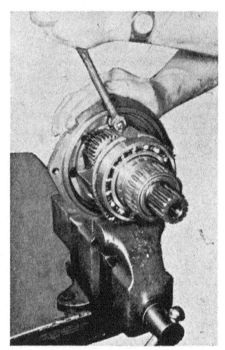

Fig. IH1143 — Removing the special over-running clutch screws.

PLANET GEARS, SUN GEARS AND OVER-RUNNING CLUTCH

201. **R&R AND OVERHAUL.** To overhaul the torque amplifier gear set and over-running clutch, first remove the TA clutch cover assembly, lined plate and clutch carrier as outlined in paragraph 200 and proceed as follows: Remove all other parts attached to clutch housing. Remove the cover plate from bottom of clutch housing. Support rear half of tractor under rear frame and attach a chain hoist around clutch housing. Remove top bolt on each side of clutch housing and install aligning dowels. Remove remaining bolts retaining clutch housing to transmission case and separate the units. Note: Lower center bolt connecting clutch housing to main frame is accessible through the lower cover plate opening.

Unbolt the transmission drive shaft bearing cage from clutch housing and withdraw the complete TA unit. Remove the small retainer ring (70—Fig. IH1139 & IH1142) from each of the four special over-running clutch screws (71—Fig. IH1139). Clamp the complete unit in a soft jawed vise and remove the four cap screws as shown in Fig. IH1143. Note: A cutout is provided in the planet carrier for this purpose. Separate the transmission drive shaft and bearing cage assembly from planet carrier (78—Fig. IH1139). Remove snap ring (63) from front of

transmission drive shaft and press the transmission drive shaft rearward out of bearing and cage. Bearing (64) can be inspected and/or renewed at this time. Inspect ramp (59), springs (68) and rollers of over running clutch. Renew damaged parts. Using OTC bearing puller attachment 952-A or equivalent, press front and rear bearings (53 & 58) from the planet carrier. It is important, when removing the front bearing to use a piece of pipe and press against the planet carrier and **not** against the primary sun gear shaft. Using a small punch and hammer, drive out the Esna roll pins retaining the planetary gear shafts in the planet carrier. Refer to Fig. IH-1145. Using OTC dummy shaft No. ED3259, push out the planet gear shafts and lift gears with rollers and dummy shaft out of the planet carrier. Be careful not to lose or damage the thrust plates (76—Fig. IH1139) as they are withdrawn. After the three compound planetary gears are removed, the primary sun gear and shaft can be withdrawn from the planet carrier.

Inspect splines, oil seal surface, bearing areas, pilot bearing and sun gear teeth of the primary sun gear and shaft for excessive wear or damage. If only the pilot bearing (49) is damaged, renew the bearing. If any other damage is found, renew the complete unit which includes an installed pilot bearing. Note: When in-

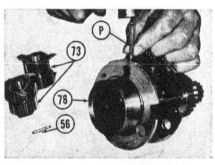

Fig. IH1145 — Using a punch and hammer to drive out the roll pins which retain the planet gear shafts in the planet carrier.

P. Punch 73. Planet gears
56. Roll pins 78. Planet carrier

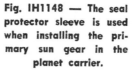

X

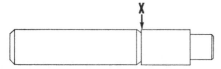

Fig. IH1146—IH tool number FES 10-29. Press pilot bearing in the primary sun gear until surface (X) is even with rear edge of the primary sun gear.

Fig. IH1148 — The seal protector sleeve is used when installing the primary sun gear in the planet carrier.

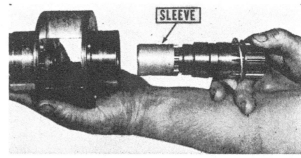

SLEEVE

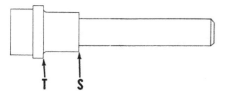

T　　S

Fig. IH1147—OTC tool number ED-3250 is used to install needle bearings in the planet carrier. Refer to text.

stalling a new pilot bearing, use IH tool No. FES 10-29 (shown in Fig. IH1146) or equivalent and press the bearing in until bearing is installed to a depth of 1.390-1.410. If bearing is installed too deep, bearing will be damaged by counterbore of primary sun gear. If bearing is not deep enough, the bearing cage will interfere with pilot radius of transmission drive shaft and secondary sun gear. Check bearing installation as follows: Temporarily install transmission input shaft and secondary sun gear in the primary sun gear without thrust washer (50—Fig. IH1139) and measure the distance between end of secondary sun gear and primary sun gear. If this distance is less than 0.080, bearing installation can be considered satisfactory. If distance is more than 0.080, interference between pilot of secondary sun gear and bearing cage is indicated and bearing installation must be rechecked. If after rechecking bearing installation the above clearance is still more than 0.080, renew the transmission drive shaft and secondary sun gear. Separate shafts and proceed as in following paragraph.

Inspect the planet carrier for rough oil seal surface, worn over-running clutch roller surface and elongated planet gear shaft holes.

Inspect the primary sun gear shaft needle bearings (55—Fig. IH1139) and the shaft oil seal (54). If bearings and/or seal are damaged, and planet gear carrier is O. K., drive out the faulty parts with a brass drift and install new bearings using OTC driving collar ED-3250 (shown in Fig. IH-1147). Press rear bearing in from front until surface (S) is even with front of planet carrier. Press the front

needle bearing in from front until surface (T) is even with front of planet carrier. Install oil seal (54—Fig. IH1139) with lip toward rear until front of oil seal is even with front of planet carrier.

Install snap ring (51) on the primary sun gear shaft and thrust washer (50) immediately ahead of the snap ring. Using OTC oil seal protector sleeve No. ED-3245, install the primary sun gear and shaft in the planet carrier as shown in Fig. IH1148.

Inspect teeth of planet gears for wear or other damage. If any one of the three gears is damaged, renew all three gears which are available in a matched set only. These planetary gears are manufactured in matched sets so the gears will have an equal amount of backlash when installed and no one gear will carry more than its share of the load. Note: The International Harvester Co. specifies that when the planet gears are removed, the planet gear shafts and needle rollers should always be renewed. The planet gear shafts are available in sets of three and the needle rollers are available in sets of 138.

Using chassis lubricant and OTC dummy shaft ED-3259, install twenty-three new needle bearings in one end of a planet gear. Slide dummy shaft into bearings and install bearing spacer (77—Fig. IH1139). With the aid of chassis lubricant, install twenty-three new needle bearings in the other end of the planet gear and slide the dummy shaft completely into the gear, thereby holding the needle bearings and spacer in the proper position. Assemble one thrust plate (76) to each end of the planet gear and install planet gear, dummy shaft and thrust plates assembly in the planet carrier. Using one of the three new planet gear shafts, push out the dummy shaft and install the Esna roll pin securing the planet gear shaft in the planet carrier. Then, observe the rear face of the planet carrier at each planet gear location where timing marks will be found. The timing marks

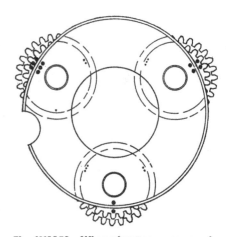

Fig. IH1149—Chassis lubricant facilitates installation of needle bearings in the planetary gear.

Fig. IH1150—When planetary gears are installed properly, punch marks on gears and planet carrier will be in register.

are punched dots (Refer to Fig. IH1150). One location has one punch mark, another location has two punch marks and the other location has three punched marks. Turn the primary sun gear shaft until the timing mark on the rear face of the installed planet gear are in register with a similar mark on the planet carrier. Now assemble the other planet gears, needle bearings, spacers, thrust plates and dummy shaft and install them so that timing marks are in register. When all three planet gears are installed properly, the single punch

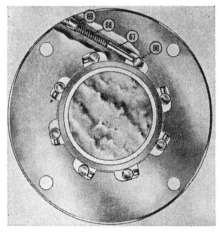

Fig. IH1151—Cut-away view shows proper installation of over-running clutch rollers (60), pins (67), springs (68) and plugs (69) in ramp.

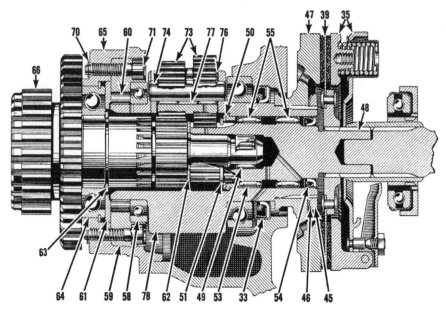

Fig. IH1152—Typical sectional view of assembled torque amplifer unit. Bearings in the unit are non-adjustable.

33. Oil seal	54. Oil seal	66. Transmission drive
35. Clutch cover assembly	55. Needle bearings	shaft and secondary
39. Driven plate	58. Bearing	sun gear
45. Nut	59. Over-running clutch	70. Retaining ring
46. Lock washer	ramp	71. Special screw
47. Clutch carrier	60. Clutch roller	73. Planetary gear
48. Primary sun gear	61. Thrust washer	74. Gear shaft
49. Roller bearing	62. Thrust washer	76. Thrust plate
50. Thrust washer	63. Snap ring	77. Bearing spacer
51. Snap ring	64. Bearing	78. Planet carrier
53. Bearing	65. Transmission drive	
	shaft bearing cage	

mark on planet carrier will be in register with single punch mark on one of the planet gears, double punch marks on carrier will register with double punch marks on one of the gears and triple punch marks on carrier will be in register with triple punch marks on one of the gears as shown in Fig. IH1150.

Install thrust washer (61B—Fig. IH-1139) on planet carrier. Assemble the pins (67—Fig. IH1151), springs (68) and rubber plugs (69) into the over-running clutch ramp and place ramp on the planet carrier. Using a small screw driver, push pins (67) back and drop rollers (60) in place as shown. Install bearing (64—Fig. IH1152) with snap ring in the rear bearing cage (65), press the transmission drive shaft (66) into position and install snap ring (63).

Place the planet carrier on the bench with rear end up and lay thrust washer (62) on the primary sun gear. Place the over-running clutch thrust washer (61 or 61A) on the ramp so that polished surface of thrust washer will contact rollers. Install the assembled transmission drive shaft and bearing cage and secure in position with the four special screws (71). After the cap screws are tightened to a torque of 40 ft.-lbs., install the small retainer rings as shown in Fig. IH1142.

Inspect the large oil seal (33—Fig. IH1152) in the clutch housing and renew if damaged. Lip of seal goes toward rear of tractor.

Using OTC oil seal protector sleeve No. ED-3253 over splines of planet carrier, insert the assembled TA unit and tighten the transmission drive shaft bearing cage cap screws securely. Assemble the remaining parts by reversing the disassembly procedure.

DIRECTION REVERSER

International 460-606-2606

International 460, 606 and 2606 tractors are available with a selective, sliding spur gear reversing unit located between the regular 5-speed transmission and the engine clutch. The unit is controlled by a hand lever which moves the sliding gear to the forward, neutral or reverse position. Refer to schematic views in Fig. IH-1154 which show the principles of operation.

205. R&R AND OVERHAUL. To remove the direction reversing mechanism, first remove the steering gear and fuel tank assembly as in paragraph 23, 65 or 200. On series 460 tractors so equipped, remove the hydraulic system control valves from the top of the clutch housing. On all models, remove the clutch housing top cover. Split the engine from the clutch housing as in paragraph 191 or 191A and the clutch housing from the transmission as in paragraph 213 or 214. Remove the engine clutch release shaft, bearing and fork. On models so equipped, remove the independent PTO drive shaft (48—Fig. IH1156), bearing cage (51),

driven gear and shaft. On all models, remove the direction reverser shifter fork (10) and shaft (8). Remove snap ring (39) and remove the clutch shaft (38). Forward end of shaft is tapped to accommodate a puller. Unbolt the collar (35) from the coupling (34) and remove the collar. Remove cap screw (32), drive shaft retainer (31) and coupling (34). Remove the four retaining cap screws and withdraw the bearing cage, planet carrier and transmission drive shaft assembly out through the rear of clutch housing. The remainder of the disassembly procedure will be evident after an examination of the unit. Refer to Fig. IH1156.

Check all parts for visible damage and renew all questionable parts. Refer to the data which follows:

Poppet balls (36) diameter........$\frac{5}{16}$"
Poppet spring (37)
    Free length ....1.086"
    Pounds test @ length ..10.2 @ 0.90"
Backlash
    Between planets
      22 & 23).........0.0010-0.0034".
    Between planets (23) and
      driven sun gear
      (11) ....0.0010-0.0034"

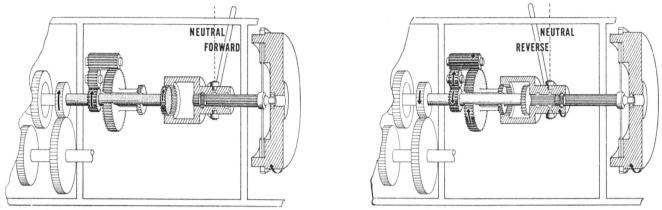

Fig. IH1154—Schematic views showing operation of direction reverser. View at left shows power flow with direction reverser in forward travel position. View at right shows power flow with direction reverser in reverse travel position.

Bore for needle bearing in
planets (22 & 23)
Diameter ....0.7834-0.7839"
Planet gear shafts (18)
Diameter ....0.5952-0.5955"
Length ....3.210"

Combined wear of shaft (18)
OD and gear (22 or 23) ID
should not exceed...........0.003"

Check the clutch housing to see if there is an oil drain hole in the position shown in Fig. IH1157.

Reassemble and install in reverse of the removal and disassembly procedure. Refer to Figs. IH1156 and IH1158. Cap screw (32—Fig. IH1156) should be torqued to 52-59 Ft.-Lbs.

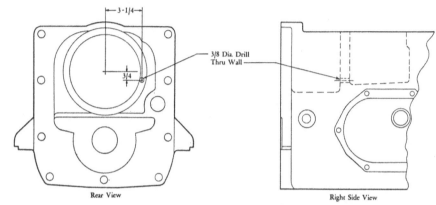

Fig. IH1157—If the clutch housing is not already so equipped, a ⅜-inch diameter oil drain hole should be drilled through the housing wall between the spline shifter coupling and the planet carrier compartments as shown.

1. Bracket
2. Grease fitting
3. Snap ring
4. Handle
6. Rod
7. Clevis
8. Shaft
9. "O" rings
10. Fork
11. Transmission drive shaft
12. Bearing carrier
13. Ball bearing
14. Snap ring
15. Spacer
16. Needle bearing
17. Planet carrier
18. Planet gear shafts (6 used)
19. Thrust plates (6 used)
20. Rollers (207 used)
21. Spacers (3 used)
22. Long planet gears (3 used)
23. Short planet gear (3 used)
24. Spacer (3 used)
25. Snap ring
26. Ball bearing
27. Bearing cage
28. Needle bearing
29. Reverse sun gear
30. Coupling
31. Transmission drive shaft retainer
32. Cap screw
33. Lock plate
34. Shifter coupling
35. Collar
36. Poppet balls (2 used)
37. Spring
38. Clutch shaft
39. Snap ring
40. Bearing cage
41. Ball bearing
42. Snap ring
43. Oil seal
44. Oil seal carrier
45. Ball (3/16-dia.)
46. Snap ring
47. Oil seal
48. IPTO drive shaft
49. Needle bearing
50. "O" ring
51. Bearing cage
52. Gasket
53. Ball bearing
54. Snap ring
55. Oil seal
56. Bearing retainer

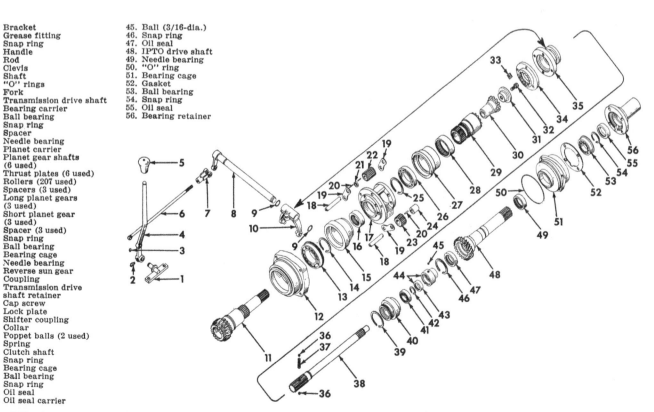

Fig. IH1156—Exploded view of the direction reverser assembly. Planet gears (22) and (23) are available in matched sets of three.

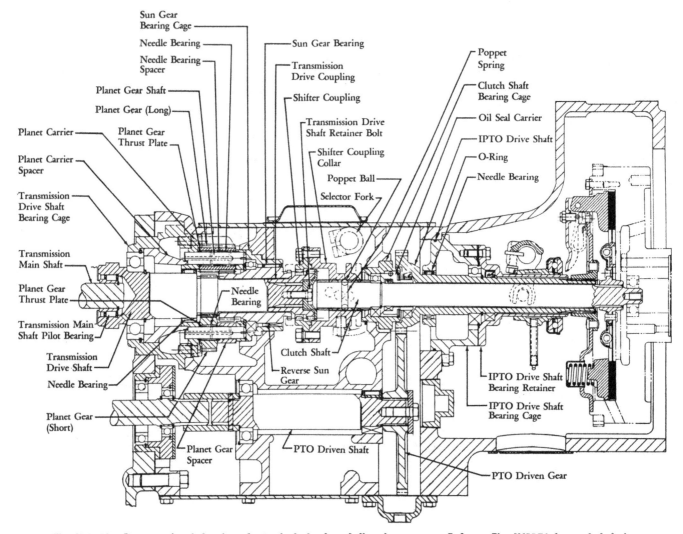

Fig. IH1158—Cross sectional drawing of a typical clutch and direction reverser. Refer to Fig. IH1156 for exploded view.

TRANSMISSION

The transmission, differential and final drive gears are all contained in the same case which is called the rear frame. A wall in the case separates the bull gear and differential compartment from the transmission gear set. Shifter rails and forks are mounted on underside of the rear frame (transmission) cover.

TOP COVER

All Models Except 460-606-2606

International

210. To remove the transmission top cover from all tractors except 460, 606 and 2606 International, first remove the hoods, fuel tank and air cleaner pipe. On Farmall tractors, disconnect the steering shaft center universal joint. On 560 and 660 International tractors, remove the steering gear assembly as outlined in paragraph 29.

On all models so equipped, disconnect the radiator shutter control rod and the follow-up cable from the control valve arm and the hydraulic lines from the rockshaft control cylinder. On all models, drain the cooling system and disconnect the tachometer cable, heat indicator sending unit, oil pressure gage line, wiring harness and controls from engine and accessories.

Unbolt the seat bracket from transmission top cover and remove the gear shift lever. Remove the cap screws which attach the fuel tank bracket to the clutch housing and the hydraulic control valve support to the transmission top cover. Attach a hoist to the fuel tank bracket, hydraulic control valves and seat assembly in a suitable manner; then, lift the complete assembly from the tractor (refer to Fig. IH1162). The top cover can be lifted off using a suitably attached hoist

after the retaining cap screws are removed.

When reinstalling the top cover, it is advisable to have the side mounted belt pulley or the cover plate removed from the front end of the cover. The shifter forks can be more easily aligned working through the large opening.

International 460

211. To remove the transmission top cover from 460 International tractors, remove the gearshift lever and seat assembly; then unbolt and lift the cover from the tractor.

NOTE: The additional work of removing or disconnecting hydraulic lines, hydraulic control valve linkage and the rockshaft may be necessary in some cases depending upon the equipment or combination of equipment that the tractor is provided with.

Fig. IH1162—To remove the transmission top cover on all except 460, 606 and 2606 International tractors, it's necessary to remove the fuel tank bracket, hydraulic control valves and seat assembly. A 560 Farmall tractor is shown.

Fig. IH1164—Before the tractor halves are rejoined, a rubber band should be positioned around the rollers of the transmission main shaft front bearing to hold the rollers in position and keep them from being pushed out of the bearing.

International 606-2606

211A. To remove the transmission top cover, along with the hydraulic lift housing, proceed as follows: Remove seat and drain lift housing. If additional working room is desired, remove both fenders. Remove both platforms. Disconnect pto control rod at pto and remove pto shifter lever and bracket. Disconnect lift links from rockshaft arms, and if so equipped, disconnect the break-away coupling bracket. Disconnect draft control rod from bellcrank, unbolt torsion bar and remove torsion bar and bellcrank. Remove torsion bar bracket. Mark the cap screw holes used and remove the quadrant. Disconnect rear junction block bracket, remove banjo bolts, then remove junction block and lines. Remove banjo bolts from front port lines and remove lines. Remove control valve through bolts and remove control valves and levers and transfer block. Disconnect lines from front of hydraulic lift housing. Unbolt rear frame top cover from rear frame, attach hoist and lift assembly from tractor.

NOTE: It is not required that control valves be removed but most mechanics prefer to do so in order to provide working room.

TRACTOR SPLIT

All Models Except 460-606-2606 International

212. The transmission may be split (detached) from the clutch housing by detaching the hydraulic support and seat from the transmission top cover, raising same slightly and moving the transmission rearward; however, some mechanics prefer to proceed as follows:

Drain the transmission and torque amplifier housings. Remove hoods and fuel tank. On Farmall tractors, disconnect the steering shaft center universal joint and unbolt the steering shaft bracket from the clutch housing. On models so equipped, disconnect the radiator shutter control rod, torque amplifier control rod and the tachometer cable. On models equipped with power steering, disconnect the return line at the rear end. On all models, drain the cooling system and disconnect the air inlet pipe, heat indicator sending unit, fuel lines, oil pressure gage line, wiring harness and controls from the engine and accessories. Disconnect the clutch rod from clutch release shaft lever, wires from the starter and the pressure hose from the left hand hydraulic control valve mounting column. Remove the cap screws which attach the fuel tank bracket to the clutch housing and the hydraulic control valve support bracket to the transmission top cover. Attach a hoist in a suitable manner and remove the complete fuel tank bracket, hydraulic control valves and seat assembly from the tractor. Support both halves of the tractor in a suitable manner, remove the cap screws which attach the clutch housing to the transmission case and separate the tractor halves.

Always renew the sealing ring (S—Fig. IH1164) as it seals the power steering and hydraulic pump suction tube. The gasket between the transmission case and the clutch housing should not cover the oil passages (O).

Before the tractor halves are rejoined, a rubber band should be positioned around the rollers of the trans-

mission main shaft front bearing as shown in Fig. IH1164. The use of a rubber band will generally serve to hold the bearing rollers in position; however, care should be exercised when joining the tractor halves to prevent a roller (or rollers) from being pushed out of the bearing.

460 International

213. To detach (split) the transmission from the clutch housing on 460 International tractors, drain the transmission and torque amplifier housings. Remove the hoods and the hydraulic control valves cover. Disconnect or remove all hydraulic tubes, wires, control rods and control cables that would hinder the separation of the tractor halves. Support the tractor halves, remove the cap screws attaching the clutch housing to the transmission case; then, separate the tractor halves.

Always renew the sealing ring (S—Fig. IH1164) as it seals the power steering and hydraulic pump suction tube. The gasket between the transmission case and the clutch housing should not cover the oil passages (O).

Before the tractor halves are rejoined, a rubber band should be positioned around the rollers of the transmission main shaft front bearing as shown in Fig. IH1164. The use of a rubber band will generally serve to hold the bearing rollers in position; however, care should be exercised when joining the tractor halves to prevent a roller (or rollers) from being pushed out of the bearing.

606-2606 International

214. To detach (split) the transmission from the clutch housing on 606 and 2606 International tractors, drain transmission and torque amplifier

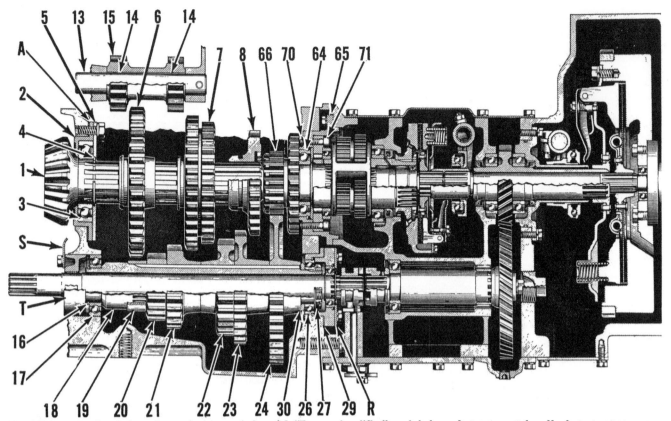

Fig. IH1165—Sectional view of a typical transmission with "Torque Amplifier" and independent power take-off. Late tractors use a roller bearing in place of ball bearing (3). The seasonal disconnect has also been discontinued. For transmission parts (items 1 through 29), refer to Fig. IH1168 for legend. For "Torque Amplifier parts (items 64 through 71), refer to Fig. IH1139 for legend.

housings. Disconnect return line from oil cooler relief valve tee. Disconnect return line at reservoir and flow control valve and remove line. If so equipped, disconnect remote lines above filter plate, loosen retaining clips and banjo bolts and swing lines out of the way. Disconnect clutch rod. Disconnect power steering line from pump flange. Loosen hydraulic lift line clips and disconnect line from flexible hose. Remove left step plate, disconnect torque amplifier linkage, then remove torque amplifier lever and bracket. Disconnect tail light wire and remove conduit from retaining clips. Remove right step plate front bracket. Support both sections of tractor, remove retaining bolts and separate tractor.

Always renew the sealing ring (S—Fig. IH1164) as it seals the power steering and hydraulic pump suction tube. The gasket between the transmission case and the clutch housing should not cover the oil passages (O).

Before the tractor halves are joined, a rubber band should be positioned around the rollers of the transmission main shaft front bearing as shown in Fig. IH1164. The use of a rubber band will generally serve to hold the bearing rollers in position; however, care should be exercised when joining the

tractor halves to prevent a roller (or rollers) from being pushed out of the bearing. Use guide studs in top hole on each side of transmission housing.

OVERHAUL

All Models

Data on overhauling the various transmission components are outlined in the following paragraphs. In general, the following paragraphs apply to all models; however, in some cases, the overhaul procedures differ on models equipped with and without torque amplifier or direction reverser as well as those equipped with and without independent power take-off. Where these differences are encountered, they will be mentioned.

215. SHIFTER RAILS AND FORKS. Shifter rails and forks are retained to bottom side of the rear frame (transmission) cover and are accessible for overhaul after removing the cover as outlined in paragraph 210, 211 or 211A. The overhaul procedure is conventional and evident after an examination of the unit.

216. TRANSMISSION DRIVING SHAFT. On models with torque amplifier, the transmission driving gear and shaft (66—Fig. IH1165) is integral with the torque amplifier secondary

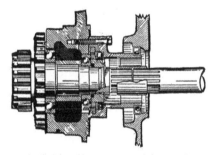

Fig. IH1166—On tractors without "Torque Amplifier" or direction reverser the area shown is different from Fig. IH1165.

sun gear and is normally serviced in conjunction with overhauling the torque amplifier unit as outlined in paragraph 201. On models with direction reverser, refer to paragraph 205.

On models without torque amplifier, the transmission driving shaft is considered a part of the transmission. To remove the drive shaft, it is first necessary to detach (split) the transmission housing from the clutch housing as outlined in paragraph 212, 213 or 214.

With the transmission detached from the clutch housing, the procedure for removing and overhauling the drive shaft is evident after an examination of the unit and reference to Figs. IH-1166, IH1167 and IH1168.

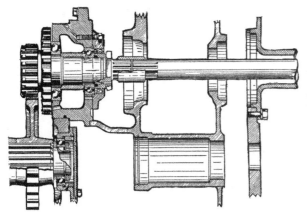

Fig. IH1167 — On tractors without "Torque Amplifier" or direction reverser, power take-off, hydraulic lift, and power steering; the area shown is different from Fig. IH-1165.

Fig. IH1169—Using a puller to remove pilot bearing from front of transmission main shaft.

217. MAINSHAFT PILOT BEARING. To remove the mainshaft pilot bearing (9 — Fig. IH1168), detach (split) the transmission housing from the clutch housing as outlined in paragraph 212, 213 or 214. Remove the bearing retaining snap ring (11) from front end of mainshaft and using OTC bearing puller or equivalent as shown in Fig. IH1169, remove the pilot bearing.

When installing the pilot bearing, the chamfered end of the inner race should be towards the forward end of the shaft.

218. MAINSHAFT (SLIDING GEAR OR BEVEL PINION SHAFT). To remove the transmission mainshaft, first detach (split) transmission from clutch housing as in paragraph 212, 213 or 214 and remove the transmission top cover.

Remove the three cap screws retaining the mainshaft rear bearing retainer (5 — Fig. IH1168) to the main case

dividing wall, move the mainshaft assembly forward and withdraw the unit, rear end first, as shown in Fig. IH1170.

Remove the mainshaft pilot bearing (9—Fig. IH1168) and slide gears from shaft. Remove snap ring (4) and press or pull rear bearing and retainer from shaft. When reassembling early models using a ball bearing on rear of mainshaft, install bearing (3) so that ball loading grooves are toward front or away from the bevel pinion gear. Later tractors use a straight roller bearing on rear of mainshaft and bearing should be installed with largest radius of inside race toward pinion gear. Use Figs. IH1165 and IH-1168 as a guide when installing the sliding gears and if the same mainshaft is installed, be sure to use the same shims (A) as were removed. If a new mainshaft is being installed, use the same shims (A) as a starting point, but be sure to check and adjust if necessary the main drive bevel gear mesh position as outlined in the main drive

bevel gear section.

219. COUNTERSHAFT. To remove the transmission countershaft, first remove the mainshaft as outlined in paragraph 218 and proceed as follows: On models with independent power take-off, remove the four cap screws retaining the independent power take-off extension shaft front bearing cage to main frame and withdraw the extension shaft, bearing cage and retainer as shown in Fig. IH1171. Working in the bull gear compartment of the main frame, remove the independent power take-off coupling shaft and the extension shaft rear bearing carrier retainer strap (S—Fig. IH1165).

On models without power take-off, remove the cap (46—Fig. IH1168).

On all models, remove the nut from forward end of countershaft and while bucking up the countershaft gears, bump the countershaft rearward until

NOT USED WITH T.A.

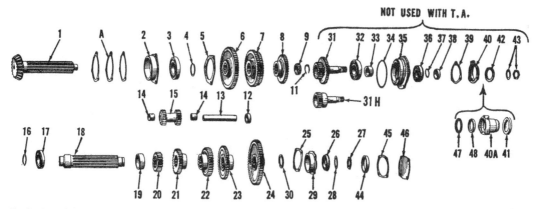

Fig. IH1168—Typical exploded view of transmission shafts, gears and associated parts. Main drive bevel pinion position is controlled by shims (A). Drive shaft (31H) is used on 460, 606 and 2606 International tractors. Late transmissions use a snap ring in place of spacer (19) and a roller bearing in place of ball bearing (3).

A. Shims (0.007, 0.015 and 0.030)	8. Fourth and fifth sliding gear	16. Snap ring	23. Fourth speed driving gear	32. Ball bearing	40A. Retainer, used with IPTO
1. Mainshaft and bevel pinion	9. Bearing	17. Bearing	24. Constant mesh gear	33. Spacer	41. Oil seal
2. Bearing cage	11. Snap ring	18. Countershaft	26. Bearing	34. Seal ring	43. Nut and lock washer
3. Rear bearing	12. Cup plug	19. Spacer	27. Nut	35. Bearing cage	44. Spacer
5. Bearing retainer	13. Reverse idler shaft	20. First speed driving gear	29. Bearing cage	36. Ball bearing	45. Gasket
6. First and reverse sliding gear	14. Reverse idler bushings	21. Second speed driving gear	30. Spacer	37. Seal ring	46. Cap
7. Second and third sliding gear	15. Reverse idler gear	22. Third speed driving gear	31. Transmission drive shaft	38. Spacer	47. Spacer
				39. Gasket	48. Seal
				40. Bearing retainer	

free from front bearing. Remove the countershaft front bearing and cage.

Withdraw the countershaft from rear and remove gears from above. The rear bearing can be removed from countershaft after removing snap ring (16—Fig. IH1168). When reassembling, use Fig. IH1165 as a guide and make certain that beveled edge of constant mesh gear spacer (30) is facing toward front of tractor.

220. **REVERSE IDLER.** With the countershaft removed as outlined in paragraph 219, the procedure for removing the reverse idler is evident. Bushings (14—Fig. IH1168) are renewable and should be reamed after installation, if necessary, to provide a recommended clearance of 0.003-0.005 for the idler gear shaft.

Fig. IH1170 — Removing the transmission mainshaft assembly.

NOTE: Tractors equipped with the Forward and Reverse Drive will not have the reverse idler gear and shaft installed in the transmission. Reverse operation of the tractor is obtained with the Forward and Reverse unit.

Fig. IH1171 — Removing the independent power take-off extension shaft, front bearing and cage.

MAIN DRIVE BEVEL GEARS AND DIFFERENTIAL

Early tractors were equipped with ball type differential carrier bearings, whereas later tractors are equipped with tapered roller bearings. Early tractors may have the later type bearings, differential, bull pinions, bull gears and brake shafts installed. A conversion package is available and should be installed, in cases where extensive damage to the final drive and/or differential occurs. After renewing a bevel pinion or ring gear, the gear mesh position and backlash as well as the differential carrier bearings should be adjusted as follows:

225. **BALL TYPE CARRIER BEARING ADJUSTMENT.** There is no adjustment, as such, for the ball type differential carrier bearings but it is important that the shims (B—Fig. IH-1180) located under the differential bearing carriers be varied to eliminate all end play from the differential unit without causing any binding tendency in the carrier bearings. To accurately check the adjustment, first install more than enough shims (B) under each carrier and proceed as follows: Bump the differential toward left side of tractor and make certain there is some backlash between the ring gear and pinion. Then check and record the amount of backlash. Now, bump the differential toward right side of tractor and again check and record the amount of backlash. Subtract the first backlash reading from the second; then, remove shims (B), equal to ½ the difference between the two backlash readings, from under each bearing carrier.

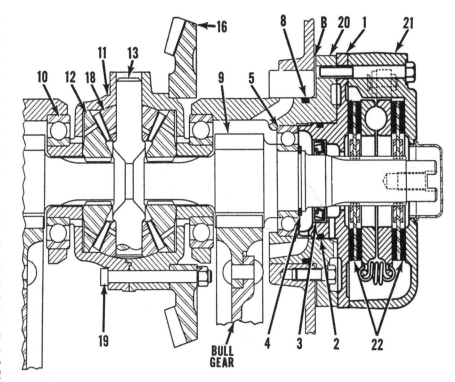

Fig. IH1180—Sectional view of early production differential and double disc brakes. Note that ball bearings (5) and (10) are used on early production models. Late tractors will have taper roller bearings at (5) and (10). Refer to Fig. IH1181 for legend.

For example, suppose the second backlash reading was 0.029 and the first was 0.015; the difference between the two readings is 0.014. In this case, remove one 0.007 thick shim (B) from each side.

Proceed to paragraphs 226 and 227 and adjust the bevel gear backlash and mesh position.

225A. **TAPERED ROLLER TYPE CARRIER BEARING ADJUSTMENT.** To adjust the taper roller bearings (9—Fig. IH1181A), first install more than enough shims (B) under each of the differential bearing carriers so that differential has a slight amount of end play and make certain there is some backlash between the ring gear

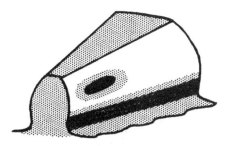

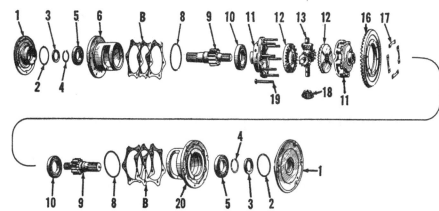

Fig. IH1180B—The correct mesh position of the bevel gears will be indicated by removal of the Prussian blue or red lead as shown by the dark spot on the drawing.

and pinion. Tighten the carrier cap screws securely. Wind a cord around large diameter of differential case, attach spring scale to end of cord and check the amount of pull required to keep the differential rolling once it has started. Record this reading which is equal to the rotational drag of the differential without any carrier bearing pre-load.

Now, remove an equal amount of shims (B) from each side until a spring scale reading of 2-7 lbs. more than that required to overcome the drag is obtained.

Proceed to paragraphs 226 and 227 and adjust the bevel gear backlash and mesh position.

NOTE: A tapered shim is used only under the right hand bull pinion bearing cage, effective on tractors at the following serial numbers: F-460, 29606; F-560, 54552; I-460, 11776; I-560, 5218; I-660, 6318; 606 and 2606, start of production. It is recommended that this tapered shim also be used on earlier tractors which have been modified to incorporate tapered roller differential bearings.

Only one tapered shim is used per tractor and if additional shims are required under the right hand bearing case, they should be positioned between the tapered shim and the final drive housing (rear frame). The tapered shim is designed to compensate for deflection of differential roller bearing when under load.

226. BACKLASH ADJUSTMENT. After the carrier bearings are adjusted as outlined in paragraph 225 or 225A, the backlash can be adjusted as follows: Transfer shims from under one bearing carrier to the other to provide 0.006-0.015 backlash between teeth of the main drive bevel pinion and ring gear. To increase backlash, remove shim or shims from carrier or ring gear side of housing and install same under carrier on opposite side. Only transfer shims, do not remove shims or the previously determined carrier bearing adjustment will be changed.

Fig. IH1181—Exploded view of the early differential, bull pinions and associated parts. Shims (B) control bearing adjustment and backlash of the main drive bevel gears. Bearings (5) and (10) may be either ball or taper roller type, see text.

B. Shims	4. Snap ring	10. Differential carrier	13. Spider
1. Bearing retainer	5. Ball bearing	bearing	16. Bevel ring gear
(inner brake plate)	6. Left bull pinion	11. Differential case	17. Lock plates
2. Seal ring	shaft bearing cage	half	18. Pinions
3. Bull pinion shaft	8. Seal ring	12. Differential side	19. Case bolts
oil seal	9. Bull pinion	gear	20. Right bull pinion
			shaft bearing cage

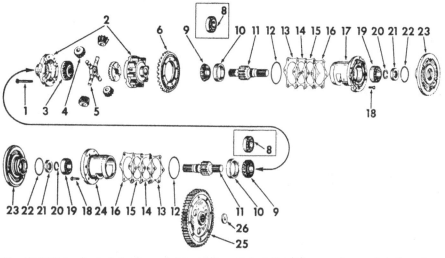

Fig. IH1181A—Exploded view of late differential, bull pinions and associated parts. Shims (13 through 16) control bearing adjustment and backlash of the main drive bevel gears. Note tapered bearings (9) which have replaced ball bearings (8) in late model tractors.

1. Case bolts	6. Bevel ring gear	12. "O" ring	22. "O" ring
2. Case halves	8. Ball bearing	13-16 Shims	23. Bearing retainer
3. Bevel gear	(not used)	17. Bearing cage (RH)	24. Bearing cage (LH)
4. Pinion gear	9. Bearing cone	19. Roller bearing	25. Bull gear
5. Spider	10. Bearing cup	20. Snap ring	26. Retainer washer
	11. Bull pinion shaft	21. Oil seal	

227. MESH POSITION. The mesh position of the bevel pinion and ring gear must be adjusted when renewing bevel pinion, bevel pinion bearings and/or ring gear. Before setting the mesh position, adjust the bearings and backlash as outlined in paragraphs 225 or 225A and 226.

The next step is to arrange shims (A—Fig. IH1165), located between the transmission mainshaft rear bearing cage and the rear frame, to provide the proper tooth contact (mesh position) of the bevel gears.

Paint the pinion teeth with Prussian

blue or red lead, rotate the ring gear by hand in normal direction of rotation and observe the contact pattern on the tooth surfaces.

The area of **heaviest contact** will be indicated by the coating being **removed** from the pinion at such points.

After obtaining the tooth contact pattern shown in Fig. IH1180B, recheck the backlash and if not within the desired limits of 0.006-0.015, adjust by transferring a shim or shims from behind one differential bearing cage to the other bearing cage until desired backlash is obtained.

RENEW BEVEL GEARS

All Models

228. To renew the main drive bevel pinion, follow the procedure outlined in paragraph 218 for overhaul of the transmission main shaft. To renew the main drive bevel ring gear, follow the procedure outlined for overhaul of the differential assembly (paragraph 229).

DIFFERENTIAL AND CARRIER BEARINGS

All Models

Differential unit is of the four pinion type mounted back of a dividing wall in the rear frame (transmission case). Refer to Fig. IH-1180 or IH1181A. The differential case halves are held together by bolts which also retain the bevel ring gear.

229. **R&R AND OVERHAUL.** To remove the differential and the main drive bevel ring gear assembly, first remove the final drive bull gears as outlined in paragraph 231 and proceed as follows: Remove brake housings, brake discs and inner brake plates Figs. IH1180, IH1181, and IH1181A. Remove the bull pinion shaft bearing cages and lift the differential and bevel ring gear assembly from tractor. Save and do not mix shims which are located between bearing cages and the rear frame. Note: The differential carrier bearings can be renewed at this time.

The procedure for overhauling the removed differential unit is evident after an examination of Fig. IH1181 or IH1181A.

When reinstalling the differential unit, assemble bull pinion shaft to left hand bearing cage and install the unit. Lower differential unit into rear frame and enter same over splines of bull pinion shaft. Install opposite (right) bearing cage and bull pinion unit.

Adjust the differential carrier bearings as outlined in paragraph 225 or 225A and check the bevel gear mesh and backlash as in paragraphs 226 and 227.

NOTE: Refer to note at end of paragraph 225A concerning the use of tapered shims used under the right hand bull pinion bearing cage.

FINAL DRIVE

As treated in this section, the final drive will include the bull pinions and integral shaft, both bull gears, axle shafts, sprocket assemblies (on high clearance models) and planetary reduction (on 660 tractors).

BULL PINION (DIFFERENTIAL) SHAFTS

All Models

230. **REMOVE AND REINSTALL.** To remove either bull pinion shaft (9—Fig. IH1180), remove the respective brake unit, remove the bull pinion shaft bearing retainer (1) (inner brake drum) and withdraw the bull pinion shaft and bearing from bearing cage.

NOTE: Bull pinion shafts and bull gears are available with 20 degree pressure angle teeth or 25 degree pressure angle teeth. Gears with different pressure angles CAN NOT be intermixed.

Teeth can be visually checked. The 25 degree pressure angle teeth are wider at the bottom and narrower at the top than are the 20 degree pressure angle teeth. See Fig. IH1181B.

BULL GEAR

All Models

231. **REMOVE AND REINSTALL.** To remove either bull gear, first remove the transmission top cover as in paragraph 210, 211 or 211A. Remove the rear wheel and cap screw (M—Fig. IH1182 or IH1183) retaining bull gear to inner end of wheel axle or bull sprocket shaft.

Some tractors may have the bull gear secured to the axle shaft by a snap ring instead of the cap screw (M—Fig. IH1182 and IH1183). Remove cap screws retaining rear axle or bull sprocket housing or sleeve to transmission housing and withdraw carrier and shaft as a unit from the rear frame. (On high clearance models withdraw carrier and one final drive unit as an assembly.) Bull gear can now be removed from rear frame (transmission housing).

NOTE: Bull pinion shafts and bull gears are available with 20 degree pressure angle teeth or 25 degree pressure angle teeth. Gears with different pressure angles CAN NOT be intermixed.

Teeth can be visually checked. The 25 degree pressure angle teeth are wider at the bottom and narrower at the top than are the 20 degree pressure angle teeth. See Fig. IH1181B.

When installing bull gear in series 660 which have gear retained by a snap ring, shims are available to reduce side play of gear on axle shaft to a minimum. Position shims between gear and snap ring.

WHEEL AXLE SHAFT

Series 460-560-606-2606 Except High Clearance and Cotton Picker

232. **REMOVE AND REINSTALL.** The wheel axle shaft can generally be removed by removing the pto rear unit (or cover) and removing the bull gear retaining cap screw and washer (or snap ring) by working through the rear opening. Remove the cap screws which attach the rear axle housing (sleeve) to the transmission housing and separate the axle and housing from the transmission housing.

NOTE: In cases where the bull gear is stuck on the axle shaft it may be necessary to remove the transmission top cover in order to work the bull gear loose.

With the axle shaft and carrier unit off tractor, remove outer bearing retainer and remove shaft from carrier by bumping on its inner end. Refer to Fig. IH1182.

Late production series 460 and 560 tractors as well as all 606 and 2606 tractors are equipped with a roller type inner axle bearing, which should be installed with the larger diameter of the inner race towards the wheel. Earlier models may have the roller bearing installed.

High Clearance Models

233. **REMOVE AND REINSTALL.** To remove wheel axle shaft or sprocket on high clearance models, proceed as follows: After draining the lubricant, remove housing pan (31—Fig. IH1183) and the connecting link from drive chain (25). It is not necessary to remove the drive chain; if however, removal of drive chain is desired, fasten a piece of flexible wire to one end and thread chain off of sprocket (20). The wire is used to help install and thread the chain over the top sprocket.

Fig. IH1181B—When renewing bull pinion shaft or bull gear, be sure gears have the same tooth angle. Refer to text.

20-degree pressure angle teeth.

25-degree pressure angle teeth.

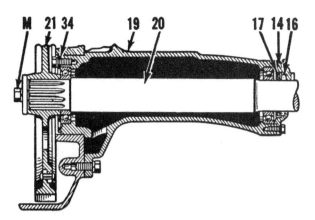

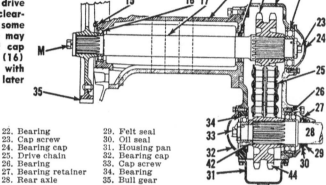

Fig. IH1182 — Sectional view of a rear axle and housing typical of all series tractors except high clearance and cotton picker models. Some models may use a snap ring to retain the bull gear on the axle shaft instead of cap screw (M) and washer. A roller bearing is now used on inner end of axle instead of ball bearing shown.

M. Cap screw
14. Bearing retainer
16. Felt seal
17. Oil seal
19. Axle housing
20. Axle shaft
21. Bull gear
34. Bearing retainer

Fig. IH1183 — Sectional view of the final drive unit used on high clearance models. On some tractors a snap ring may be used instead of cap screw (M). Bearing (16) has been replaced with a roller bearing on later models.

M. Cap screw
15. Bearing retainer
16. Bearing
17. Sprocket shaft
18. Housing carrier
19. Oil seal
20. Drive sprocket
21. Rear axle housing
22. Bearing
23. Cap screw
24. Bearing cap
25. Drive chain
26. Bearing
27. Bearing retainer
28. Rear axle
29. Felt seal
30. Oil seal
31. Housing pan
32. Bearing cap
33. Cap screw
34. Bearing
35. Bull gear

Remove inner and outer wheel axle bearing retainers (27 and 32) and inner bearing cap screw (33). Bump wheel axle shaft (28) on inner end and out of sprocket while bucking up the sprocket (44).

When reinstalling axle shaft, install spacer (42) between inner bearing (34) and sprocket hub with wide face of spacer nearest the sprocket hub.

Series 660 International

234. **REMOVE AND REINSTALL.** To remove the wheel axle shaft (26—Fig. IH1184), unbolt and remove the cover plate (36) and driver assembly (35). Withdraw the wheel axle.

UPPER SPROCKET SHAFT

High Clearance Models

235. **REMOVE AND REINSTALL.** To remove the upper sprocket shaft (17—Fig. IH1183) on high clearance models proceed as follows: Remove bull gear as outlined in paragraph 231. Remove drive chain as outlined in first part of paragraph 233. Remove outer bearing cap (24) and cap screw (23) retaining bearing (22) to shaft. Remove cap screws retaining sprocket shaft carrier (18) to transmission housing and remove carrier (18). Bump

sprocket shaft out of sprocket and withdraw the sprocket.

Late production tractors are equipped with a roller type inner axle bearing, which should be installed with the larger diameter of the inner race towards the wheel. Earlier models may have the roller bearing installed.

DRIVE HUB

Series 660 International

236. **REMOVE AND REINSTALL.** To remove and reinstall the drive hub (29—Fig. IH1184) and bearings (28 and 30), proceed as follows: Support the tractor and remove the tire, wheel and wheel hub. Unbolt and remove the cover plate (36) and driver assembly (35). Unlock and remove the bearing adjusting nut (33) and withdraw the lock washer (32) and the spacer washer (31). Bump or pull the drive hub (29) and bearing cone (30) from the wheel carrier (24).

The oil seal (25) should be installed with the larger lip of same facing toward the drive hub bearings (28 and 30).

Some early tractors were not equipped with the spacer washer (31); however, this washer should be in-

stalled on these tractors during reassembly.

Reinstall in the reverse of the removal procedure. The bearings (28 and 30) should be adjusted as follows: Install and tighten nut (33), then back the nut out until there is approximately 0.020 clearance between the nut and the lock washer (32). Pull the wheel hub out enough to move the bearing cone out. Wind a cord around the wheel hub and attach a spring scale to the end of the cord. Check the amount of pull required to rotate the hub under these conditions. Read scale while hub is rotating. Tighten the bearing adjusting nut (33) until a pull 6 to 10 pounds more than that required to overcome drag is obtained. Lock adjustment with the lock washer (32).

INNER AXLE

Series 660 International

237. **REMOVE AND REINSTALL.** To remove and reinstall the inner axle shaft (5—Fig. IH1184) and bearings, drain the oil from the transmission housing and remove the fender. Refer to paragraph 238 and remove the planetary reduction unit. Remove the pto rear unit (or cover) and remove the cap screw (or snap ring) which retains the bull gear on the axle inner end. NOTE: In some cases it may be necessary to remove the transmission top cover and remove bull gear as outlined in paragraph 231. Detach and lower the drawbar from the axle housing. Support the axle housing with a hoist, remove the cap screws which attach the axle housing to the transmission housing, then separate the axle and housing from the tractor.

Remove the inner ball bearing retaining snap ring (3—Fig. IH1184) and bump the shaft (5) and inner bearing (4) out of the housing toward the inner end. The needle bearing (7) and (seal or) seals (6) can be removed from the housing and renewed.

Reinstall in the reverse of the removal procedure. Two oil seals (6) were used in early production tractors; however, only one seal should be installed when servicing. When only one seal (6) is used, it should face toward center of tractor and breather (9) should be discarded and a pipe plug installed.

Late production tractors are equipped with a roller type inner axle bearing, which should be installed with the larger diameter of the inner race towards the wheel. Earlier models may have the roller bearing installed.

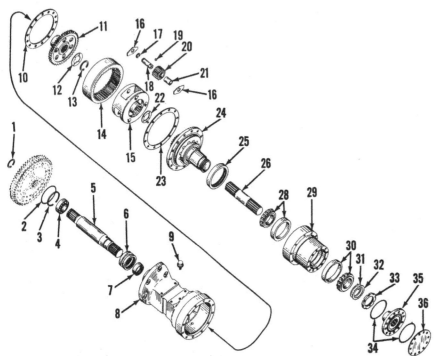

Fig. IH1184—Exploded view of the rear axles and planetary unit used on 660 International tractors. Two oil seals (6) were used in early production tractors however, only one seal should be installed when servicing. When only one seal (6) is used, it should face toward center of tractor and breather (9) should be discarded and a plug (part No. 110 891H) installed. Late models use a cap screw and washer to retain bull gear on axle instead of snap ring (1).

1. Snap ring (early models), cap screw and washer (late models)	17. Snap ring
	18. Shaft
	19. Ball
2. "O" ring	20. Planet gear
3. Snap ring	21. Bushing
4. Bearing	22. Thrust washer
5. Inner axle shaft	23. Gasket
6. Two opposed lip seals	24. Wheel carrier
	25. Oil seal
7. Needle bearing	26. Wheel axle shaft
8. Inner axle housing	28. Bearing
9. Breather	29. Drive hub
10. Retaining ring	30. Bearing
11. Ring gear hub	31. Spacer washer
12. Thrust washer	32. Lock washer
13. Snap ring	33. Bearing adjusting nut
14. Ring gear	34. "O" rings
15. Planet carrier	35. Driver assembly
16. Thrust plate	36. Cover plate

Fig. IH1187 — Exploded view of the rear axle and drive gear unit used on cotton picker equipped tractors. On later models ball bearing (1) has been replaced with a roller bearing.

1. Bearing
2. Inner axle housing
4. Inner axle shaft
5. Ball bearing
6. Oil seal
7. Washer
8. Gasket
9. Bearing cap
11. Gasket
12. Drive gear housing
13. Ball bearing
14. Drive pinion
15. Ball bearing
16. Gasket
17. Housing cover
18. Washer
19. Pin
20. Bearing cover
21. Gasket
22. Lock
23. Ball bearing

Fig. IH1185—The tire, wheel and wheel hub removed showing the wheel carrier (24) and the inner axle housing (8) on 660 International.

PLANET ASSEMBLY

Series 660 International

238. **REMOVE AND REINSTALL.** To remove the planet assembly, it is first necessary to remove the tire, wheel and wheel hub. Refer to Fig. IH1185 and remove cap screws which attach the wheel carrier (24) to the inner axle housing (8) and remove the carrier assembly. The planet carrier and gears can now be withdrawn from the inner axle housing.

The ring gear and hub can be removed after removing the snap ring (13—Fig. IH1184). The sun gear is part of the wheel carrier (24). When reinstalling, reverse removal procedure.

239. **OVERHAUL.** To overhaul the planetary assembly, first remove the unit as described in paragraph 238. The planet gears (20—Fig. IH1184) can be removed after removing the snap rings (17) and withdrawing shafts (18). NOTE: Do not lose the small balls (19) as the shafts are removed. Thrust washers (12, 16 and 22) and bushings (21) should be checked for evidence of excessive wear. The ring gear (14) is detachable from the hub (11) and can be renewed separately.

DRIVE GEAR UNIT

Cotton Picker Models

240. **R&R AND OVERHAUL.** The procedure for removing, disassembling and reinstalling the drive gear unit on these tractors will be evident after an examination of the unit and reference to Fig. IH1187.

Late production tractors are equipped with a roller type inner axle bearing, which should be installed with the larger diameter of the inner race towards the wheel. Earlier models may have the roller bearing installed.

24. Oil pan	29. Oil seal
25. Gasket	30. Felt washer
26. Driven gear	31. Gasket
27. Wheel axle shaft	32. Oil seal cover
28. Ball bearing	34. Snap ring

BRAKES

All Models

241. Brakes are of the double disc, self energizing type, which are splined to the outer ends of the bull pinion shaft. The molded linings are bonded to the brake discs. Procedure for removing the lined discs will be evident after an examination of the unit and reference to Fig. IH1193.

242. **ADJUSTMENT.** To adjust the brakes, loosen jam nut (1—Fig. IH-1194 or IH1195) and turn the adjusting nut (2) either way as required to obtain the correct pedal free travel of 1¼ inches for Farmall 460 tractors;

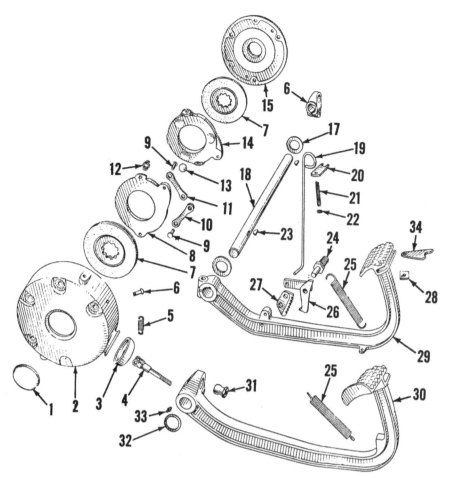

Fig. IH1193—Exploded view of a typical brake and associated parts. Series 460, 606 and 2606 tractors use four balls (13); Series 560 and 660 use five balls. Earlier brakes may not include anchor (6) and brake positioning spring (5); however, if brake dragging is encountered, positioning spring can be installed. Refer to text.

1. Expansion pug	10. Actuating link (yoke)	17. Thrust washer	26. Lock pawl
2. Housing		18. Pedal shaft	27. Ratchet
3. Boot	11. Actuating link (plain)	19. Brake lock rod	28. Tension spring
4. Operating rod		20. Bracket	29. Pedal L.H.
5. Positioning spring	12. Extension spring	21. Spring	30. Pedal R.H.
6. Anchor	13. Ball	22. Washer	31. Operating rod ball
7. Brake disc	14. Actuator	23. Woodruff key	32. Washer
8. Actuator	15. Bull pinion bearing retainer	24. Brake lock shaft	33. Grease fitting
9. Stud		25. Pedal return spring	34. Pedal lock plate
	16. Lever		

should be taken between pedals and rear frame cover. Brakes can be equalized by loosening the tight brake. When adjustment is completed, tighten jam nut (1).

242A. NOTE: If problems of brake dragging or chattering should occur on earlier models of tractors, a brake positioning spring and anchor is available to insure that brakes are held against their stops when not in use.

To install brake positioning spring, proceed as follows: Disengage boot from its groove in brake housing, then using upper edge of boot groove as a reference point, measure $1\frac{7}{8}$ inches on series 460, or $1\frac{3}{4}$ inches on series 560 and 660, upward on brake housing and scribe a horizontal line on brake housing. Measure along the scribed horizontal line $\frac{9}{16}$-inch outward from inside edge of housing and center punch housings for subsequent drilling. Remove housings and drill a $\frac{3}{16}$-inch hole through housings at the center punch marks. Be sure to drill holes square with brake housings.

NOTE: While the above information will give location of holes to be drilled, it is probable that a comparison between old and new housings would be more satisfactory.

Be sure legs of cotter pin are wrapped securely around brake operating rod clevis pin, then insert positioning spring anchor in the drilled hole and attach positioning spring to anchor and through loop of cotter pin head.

Reinstall brake assemblies and adjust as outlined in paragraph 242.

BELT PULLEY

Two types of belt pulley units are used, one is a side mounted pulley (Fig. IH1196) the other is a rear mounted pulley (Fig. IH1197).

243. **REMOVE AND REINSTALL.** The procedure for removing either type of belt pulley unit, is evident after an examination of the installation.

244. **OVERHAUL.** The procedure for disassembling and reassembling the belt pulley is evident after an examination of the unit and reference to Fig. IH1196 or IH1197. Shims (15 and 24) control the mesh position and backlash of the bevel gears. Recommended bevel gear backlash is 0.008-0.010.

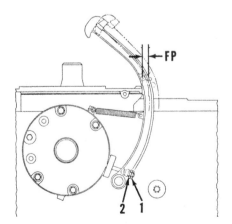

Fig. IH1194—The brake pedal free play (FP) should be adjusted by turning the adjusting nut (2). Refer to Fig. IH1195 for 460, 606 and 2606 International.

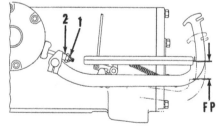

Fig. IH1195 — International 460, 606 and 2606 brake pedal free play (FP) should be adjusted by turning the adjusting nut (2).

$1\frac{1}{2}$-$1\frac{5}{8}$ inches for all 560 and 660 tractors; $2\frac{1}{4}$ inches for International 460 tractors or $1\frac{7}{8}$ inches for International 606 and 2606 tractors. Measurement

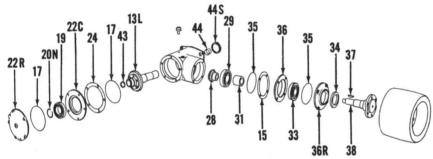

Fig. IH1196—Exploded view of the side mounted belt pulley unit used on some models. Gear (42) meshes with spur gear on the transmission drive shaft.

11. Drive shaft	19. Ball bearing	28. Driven bevel gear	35. Seal ring
12. Woodruff key	20. Nut	29. Ball bearing	36. Bearing cage
13. Bevel drive pinion	21. Gasket	30. Snap ring	37. Woodruff key
15. Shims (0.008, 0.015 and 0.030)	22. Bearing cage cover	31. Spacer	38. Pulley shaft
	24. Shims (0.008, 0.015 and 0.050)	32. Snap ring	42. Driving spur gear
17. Seal ring	26. Nut	33. Ball bearing	43. Cup plug
18. Drive shaft bearing cage		34. Oil seal	44. Drive shaft bushing

Fig. IH1197—Exploded view of rear mounted belt pulley unit used on some models. Drive is taken from the pto shaft.

13L. Drive gear and shaft	20N. Snap ring	29. Ball bearing	36R. Bearing retainer
15. Shims (0.007, 0.012 and 0.030)	22C. Bearing cage	31. Spacer	37. Woodruff key
	22R. Bearing retainer	33. Ball bearing	38. Pulley shaft
17. "O" rings	24. Shims (0.007, 0.012 and 0.030)	34. Oil seal	43. Cup plug
19. Ball bearing	28. Driven bevel gear	35. "O" ring	44. Drive shaft bushing
		36. Bearing cage	44S. Oil seal

POWER TAKE-OFF

Two types of pto rear units have been used. Early tractors were equipped with a planetary type pto as shown in Fig. IH-1202. Later tractors were equipped with a clutch type pto as shown in Fig. IH1213. Late clutch type pto units were modified in that an adjusting hole was provided so that unit could be adjusted while still on the tractor. A special adjusting tool is furnished with each tractor.

ADJUSTMENTS

Planetary Type

245. REACTOR BANDS. To adjust the reactor bands, remove the adjusting screw cover, loosen lock nuts (N—Fig. IH1198) and back off the adjusting screws (M) approximately four turns. Hold the operating lever so that the pawl on the control lever is cen-

tered in the middle land of the quadrant in a manner similar to that shown at (X).

Turn the adjusting screws in (clockwise) until screws are reasonably tight and set the lock nuts finger tight. Move the operating lever back and forth several times; then, back-off the adjusting screws approximately one turn and tighten the lock nuts.

Using a $\frac{5}{16}$-inch rod approximately 8 inches long as a lever in hole of pto shaft, check to make certain that shaft is free to turn only when the pawl is exactly centered on the middle land of the quadrant as shown at (X—Fig. IH1198). The shaft should not turn with lever in any other position.

If shaft turns freely with the control lever in any other position the adjusting screws are not evenly adjusted and

Fig. IH1198 — When adjusting the early type power take-off reactor bands, the pawl on the handle should be centered on the land as shown at (X).

readjustment is necessary.

If shaft is not free to turn in any position of the lever, turn the adjusting screws out (counter-clockwise) one-half turn, lock and recheck.

Tighten lock nuts when adjustment is complete.

Clutch Type (Early)

246. OVERCENTER CLUTCH. To adjust the overcenter clutch, the pto rear unit must be removed from tractor. With unit removed, remove the two cap screws which retain the two halves of the unit together. Place rear cover and clutch assembly in a vise with the clutch assembly upward. Place a 1¼ inch pipe at least two feet long over the cross shaft lever to provide leverage to operate clutch and adjust clutch as follows: Disengage clutch, if necessary. Insert the blade of a screwdriver under lock plate (36—Fig. IH1200), pry lock plate up to disengage the lock pin from its groove (notch) in spider (35), then turn lock plate slightly so it will engage the thread relief and hold pin in the disengaged position. Turn spider toward pressure plate (tighten) one locking position at a time and at each locking position, engage clutch to check that it goes overcenter. Continue this operation until the point is reached where the clutch will no longer go overcenter. Now back-off spider one locking position and engage lock pin. Clutch is correctly adjusted at this time and when engaged, should go overcenter with a distinct snap.

After adjusting pto clutch, a brake (anti-creep) adjustment must also be made. Refer to paragraph 248.

Clutch Type (Late)

247. OVERCENTER CLUTCH. On late model clutch type pto, the over-

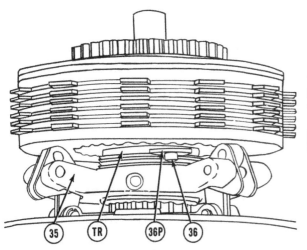

Fig. IH1201—The linkage should be correctly adjusted as described in text.

CL. Clevis   R. Rod

center clutch can be adjusted with the unit on the tractor as follows: Remove the large plug from top side of pto housing. Place control lever in middle (neutral) position. Rotate pto output shaft by hand until flat portion of brake plate is in line with plug hole. Now insert the special adjusting tool (furnished with tractor) into plug hole and between brake plate and housing and into slot of adjusting spider. See Fig. IH1201A. Hold front of adjusting tool downward, push tool forward

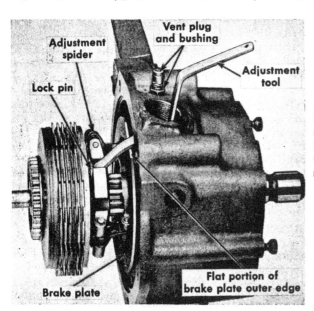

Adjustment spider

Vent plug and bushing

Adjustment tool

Lock pin

Brake plate

Flat portion of brake plate outer edge

and disengage lock pin from adjusting spider. Hold tool and lock pin in this position and turn pto output shaft counter-clockwise to tighten clutch. Adjust clutch one notch (locking position) at a time and recheck after each adjustment. Clutch should go over-center with a distinct snap. Usually one or two notches is sufficient. Over-adjusting will prevent full engagement of clutch.

After adjusting pto clutch, a brake (anti-creep) adjustment must also be made. Refer to paragraph 248.

Clutch Type (Early and Late)

248. **LINKAGE.** To adjust the linkage (brake) on clutch type pto, vary the control rod length until the distance between front of pawl and front of pawl slot in control handle quadrant is as follows: Farmall 560, International 560 and 660, ½-inch; Farmall 460, ⅜-inch; International 460, 606 and 2606, $\frac{1}{16}$-inch.

Fig. IH1201A — View showing method of adjusting the late clutch type pto. Unit is shown removed for illustrative purposes.

OVERHAUL

The occasion for overhauling the complete power take-off system will be infrequent. Usually, any failed or worn part will be so positioned that localized repairs can be accomplished. The subsequent paragraphs will be outlined on the basis of local repairs.

Planetary Type

250. **RENEW REACTOR BANDS.** To renew the reactor bands (34—Fig. IH1202), remove the pto shaft guard and the band adjusting screw cover. Unbolt bearing retainer (47) from the housing cover and remove the retainer. Unlock the pto shaft nut (45), place the operating lever in the forward position to prevent shaft from turning and remove nut (45). Remove the cap screws retaining the rear cover to the housing, place operating lever in the neutral position and turn the pto shaft to align threaded holes in the rear drum (38) with the unthreaded holes in the rear cover.

Using OTC puller ED-3262 or equivalent as shown in Fig. IH1203, remove the rear drum and cover. Loosen the band adjusting screws and remove bands. Inspect bands for distortion, lining wear and for looseness of strut pins. When reassembling, adjust the bands as outlined in paragraph 245.

251. **PLANET GEARS, SUN GEARS AND SHAFTS.** To overhaul the pto rear unit, first drain transmission, then remove the reactor bands as outlined in paragraph 250. Unbolt and remove the unit from the tractor.

With the unit on bench, remove cap screws securing bearing cage (12—Fig. IH1202 or IH1206) to housing and remove bearing cage with ring gear and shaft. Remove the front bearing retainer (16) and seal (17). Remove snap ring (14) and press the ring gear and shaft from bearing (13). The front bearing (13) can be removed from cage at this time. To remove the ring gear rear bearing (9), remove snap ring (10) and using a punch through the two holes in the ring gear hub, bump bearing from shaft. Inspect needle roller pilot bearing (7). If the bearing is damaged, it can be removed, using a suitable puller as shown in Fig. IH-1204.

Withdraw the planet carrier and pto shaft from housing and mark the rear face of each planet gear so it can be installed in the same position. Using a punch as shown in Fig. IH1205, remove the roll pins which retain the planet gear shafts in the planet carrier and remove the shafts, gears, spacers and needle bearings. Remove snap ring (26—Fig. IH1202 or IH1206) from sun

Fig. IH1202 — Exploded view of the rear section of the early (planetary) type independent power take-off. Planetary gears (4) are available in sets only.

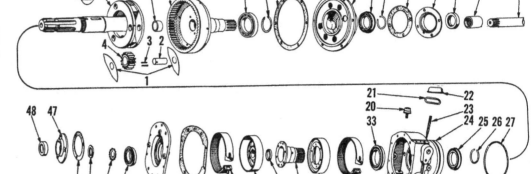

1. Thrust washers
2. Planet gear shaft
3. Needle bearing rollers (72)
4. Planet gear
5. Key
6. Planet carrier and pto shaft
7. Needle bearing
8. Ring gear and shaft
9. Drive shaft rear bearing
10. Snap ring
11. Gasket
12. Bearing cage
13. Drive shaft front bearing
14. Snap ring
15. Gasket
16. Bearing retainer

17. Oil seal
18. Coupling
19. Coupling shaft
20. Breather
21. Gasket
22. Anchor bolt cover

23. Bolt
24. Housing
25. Bearing
26. Snap ring
27. Seal ring
28. Oil seal

29. Snap ring
30. Bushing
31. Lever
32. Key
33. Bearing
34. Bands

35. Brake drum
36. Sun gear
37. Spacer
38. Creeper drum
39. Gasket
40. Housing cover

43. Bearing
44. Lock washer
45. Nut
46. Gasket
47. Bearing retainer
48. Oil seal

gear and press sun gear (36) and drum (38) from housing. Bearing (25) can be removed from housing at this time. If rear bearing (33) is damaged, use a punch through holes in hub and drift the bearing from the sun gear as shown in Fig. IH1207. Remove operating linkage from side of housing. To disassemble the spring retainer plug assembly, turn spring anchor block (57—Fig. IH1209) clockwise to relieve spring pressure and remove snap ring (56).

NOTE: Later spring retainer plug assemblies use two (inner and outer) springs.

Caution: If the anchor bolt is broken or if spring tension cannot be relieved, use care when removing the snap ring.

Inspect all parts and renew any which are excessively worn. The planet gears are available only in sets as are the planet gear needle bearings and the gear shafts. Friction surface of drums should not be excessively worn.

When reassembling, reverse the disassembly procedure. OTC dummy shaft No. ED-3258-1 is used to assemble the needle bearings in the planet gears and to install the planet gears, with thrust plates to the planet carrier. With the planet gear, thrust plates and dummy shaft in position in the planet carrier, push the dummy shaft out with the new shaft. Secure the planet gear shafts to the planet carrier with the roll pins.

When installation is complete, adjust the reactor bands as outlined in paragraph 245.

252. COUPLING SHAFT. To renew the pto coupling shaft (19—Fig. IH-1206), remove the complete pto rear unit (Fig. IH1208) and withdraw coupling shaft from rear frame.

Fig. IH1203 — Using a puller to remove cover and rear drum from the early type independent power take-off rear unit.

253. EXTENSION SHAFT. To remove the pto extension shaft (Fig. IH-1210), first detach (split) clutch housing from rear frame as outlined in paragraph 212, 213 or 214. Remove the seasonal disconnect coupling from the front of shaft (if so equipped). Unbolt the extension shaft front bearing retainer and cage from transmission and withdraw the extension shaft assembly as shown in Fig. IH1211.

Remove the coupling shaft. Remove cap screw and strap which retains the extension shaft rear bushing carrier (Fig. IH1210) in the rear frame and remove the bushing carrier. Bushing can be renewed if it is worn.

Reinstall the extension shaft by reversing the removal procedure.

254. DRIVEN SHAFT AND GEAR. Refer to Fig. IH1210. To remove the pto driven shaft and gear, first detach the transmission from the clutch housing as outlined in paragraph 212, 213 or 214 and proceed as follows:

Fig. IH1204 — Removing the needle pilot bearing from ring gear.

Remove the driven gear cover and the clutch and flywheel opening cover from bottom of clutch housing and the large pipe plug which is located directly in front of the driven shaft. Remove cap screw and washer retaining driven gear to shaft and remove snap ring from behind the driven shaft rear bearing. Withdraw the driven shaft rearward and the gear from below. NOTE: In some cases it may be necessary to detach the clutch housing from the engine and use a brass drift to bump the shaft rearward. Use two 1 x 2 inch wood blocks about six inches long between rear of pto driven gear and clutch housing to prevent damage to front needle bearing. If the driven shaft front needle bearing is damaged, it can be renewed in a conventional manner at this time. The needle bearing race on shaft can also be renewed if damaged. Remove the race with a brass drift to avoid damaging the shaft.

When reassembling, reverse the disassembly procedure and use sealing

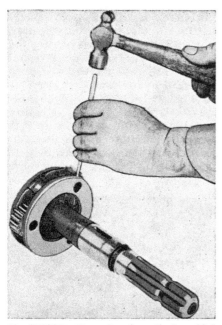

Fig. IH1205—Removing the roll pins which secure the planet gear shafts in the planet carrier.

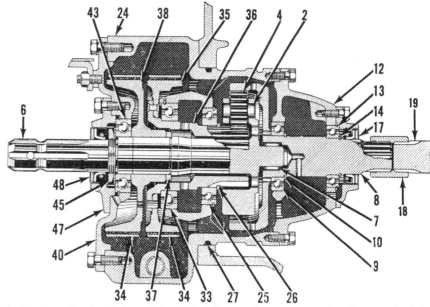

Fig. IH1206—Sectional view of the early (planetary) type power take-off rear unit. Refer to legend under Fig. IH1202.

compound around pipe plug in front of driven gear.

NOTE: Tighten the retaining cap screw to 115-130 ft.-lbs. torque.

255. DRIVE SHAFT. Refer to Fig. IH1210. To remove the driving shaft and integral gear, first detach (split) engine from clutch housing as outlined in paragraph 189, 190, 191 or 191A and remove the engine clutch release bearing and shaft. Unbolt the drive shaft front bearing cage and withdraw the drive shaft and bearing cage from clutch housing. The need and procedure for further disassembly is evident.

CLUTCH TYPE

256. REAR UNIT. To overhaul the rear pto unit, it is necessary to first remove the unit. Remove the two cap screws which attach the two halves of the rear unit together and separate the halves. Refer to Fig. IH1213. Remove the two bearing retaining screws (55) and withdraw the input shaft (49) and bearing (50) from the housing (53).

The remainder of disassembly and overhaul is obvious. Refer to the following table for information on the rear unit.

Pressure plate thickness...0.215-0.220
    Max. allowable warpage.......0.012
Internal splined disc
    thickness 0.060-0.065
    Max. allowable wear.........0.010
External splined disc
    thickness 0.110-0.115
Max. allowable wear of
    all discs 0.125

Brake plate springs
    Free length ....2.250 in.
    Test length ....1.649 in.
    Test load .... 135 lbs.
Lock pin spring
    Free length ....1.250 in.
    Test length ....1.000 in.
    Test load ....5 lbs.

Reassemble in reverse of disassembly procedure and adjust the clutch as in paragraph 246 or 247. Reinstall in reverse of the removal procedure and adjust the linkage as in paragraph 248.

257. COUPLING SHAFT. To renew the pto coupling shaft (19—Fig. IH1213), remove the complete pto rear unit (Fig. IH1212) and withdraw coupling shaft from rear frame.

258. EXTENSION SHAFT. To remove the pto extension shaft (Fig. IH1210), first detach (split) clutch housing from rear frame as outlined in paragraph 212, 213 or 214. Remove the seasonal disconnect coupling from front of shaft (if so equipped). Unbolt the extension shaft front bearing retainer and cage from transmission and withdraw the extension shaft assembly as shown in Fig. IH1211.

Remove the coupling shaft. Remove cap screw and strap which retains the extension shaft rear bushing carrier (Fig. IH1210) in the rear frame and remove the bushing carrier. Bushing can be renew if it is worn.

Reinstall the extension shaft by reversing the removal procedure.

259. DRIVEN SHAFT AND GEAR. Refer to Fig. IH1210. To remove the pto driven shaft and gear, first detach

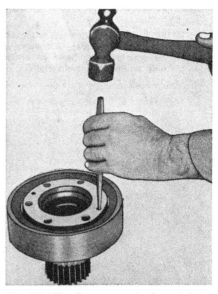

Fig. IH1207—Using a punch through holes of sun gear to remove rear bearing.

Fig. IH1208—Rear view of the tractor, showing the installation of the early (planetary) type power take-off rear unit.

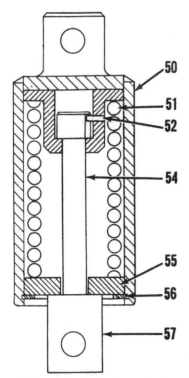

Fig. IH1209—Sectional view of spring retainer plug assembly. Refer to caution in text before disassembling. Later units also include additional inner spring.

50. Spring sleeve	55. Retainer plate
52. Roll pin	56. Snap ring
54. Anchor bolt	57. Anchor block

the transmission from the clutch housing as outlined in paragraph 212, 213 or 214 and proceed as follows:

Remove the driven gear cover and the clutch and flywheel opening cover from bottom of clutch housing and the large pipe plug which is located directly in front of the driven shaft. Remove cap screw and washer retaining driven gear to shaft and remove snap ring from behind the driven shaft rear bearing. Withdraw the driven shaft

rearward and the gear from below. NOTE: In some cases it may be necessary to detach the clutch housing from the engine and use a brass drift to bump the shaft rearward. Use two 1 x 2 inch wood blocks about six inches long between rear of pto driven gear and clutch housing to prevent damage to front needle bearing. If the driven shaft front needle bearing is damaged, it can be renewed in a conventional manner at this time. The

Fig. IH1211 — Removing the independent power take-off extension shaft from transmission housing.

Fig. IH1212 — Rear view showing the installation of the clutch type power take-off rear unit.

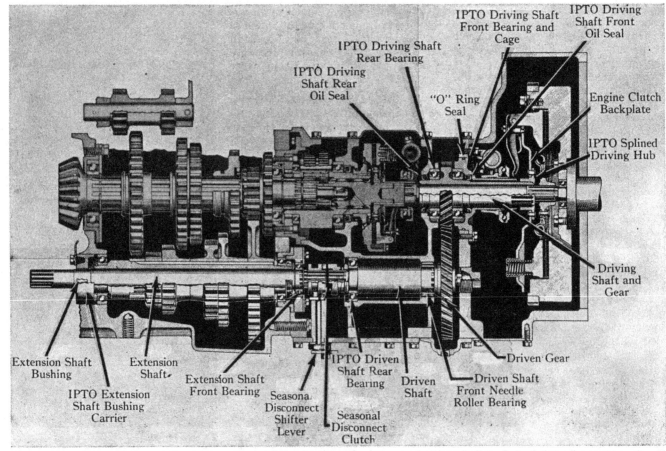

Fig. IH1210—Sectional view of the clutch and transmission housing, showing a typical installation of the independent power take-off shafts and gears. Seasonal disconnect assembly is no longer used.

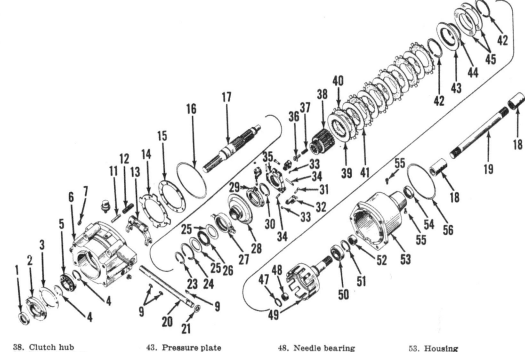

Fig. IH1213 — Exploded view of the early clutch type independent pto rear unit. Refer to Fig. IH1202 for the early (planetary) type.

1. Oil seal
2. Bearing retainer
3. Gasket
4. Snap rings
5. Ball bearing
6. Housing cover
7. Plug
9. Woodruff keys
11. Dowel pins (4 used)
12. Brake plate springs (4 used)
13. Clutch fork
14. Brake plate
15. Brake facing
16. "O" ring
17. Output shaft
18. Couplings
19. Coupling shaft
20. Cross shaft
21. Oil seal
23. Snap ring
24. Snap ring
25. Bearing races (2 used)
26. Needle thrust bearing
27. Release bearing
28. Release sleeve
29. Actuator
30. Snap ring
31. Pins (3 used)
32. Actuating lever (3 used)
33. Retainer ring (6 used)
34. Pins (3 used)
35. Spider
36. Lock pin
37. Spring

38. Clutch hub	43. Pressure plate	48. Needle bearing	53. Housing
39. Pressure plate	44. Snap ring	49. Input shaft	54. Oil seal
40. Lined discs (6 used)	45. Belleville washers	50. Ball bearing	55. Bearing retaining screws
41. Discs (5 used)	(2 used)	51. Snap ring	(2 used)
42. Snap rings	47. Snap ring	52. Needle bearing	56. "O" ring

needle bearing race on shaft can also be renewed if damaged. Remove the race with a brass drift to avoid damaging the shaft.

When reassembling, reverse the disassembly procedure and use sealing compound around pipe plug in front of driven gear.

260. **DRIVE SHAFT.** Refer to Fig. IH1210. To remove the driving shaft and integral gear, first detach (split) engine from clutch housing as outlined in paragraph 189, 190, 191 or 191A and remove the engine clutch release bearing and shaft. Unbolt the drive shaft front bearing cage and withdraw the drive shaft and bearing cage from clutch housing. The need and procedure for further disassembly is evident.

HYDRAULIC LIFT SYSTEM ("HYDRA-TOUCH")

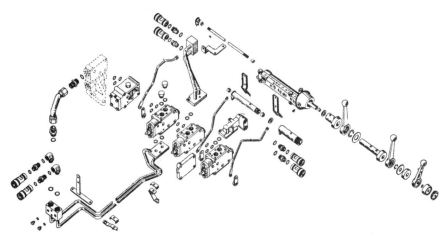

Fig. IH1215—View showing the general lay-out of component parts of a typical "Hydra-Touch" hydraulic system. View shown is for an International 460, however, other models so equipped will be similar.

The series 460, 560 and 660 tractors are equipped with a hydraulic lift system referred to as "Hydra-Touch". See Fig. IH1215. Refer to paragraph 290 for the series 606 and 2606 hydraulic lift system.

NOTE: The maintenance of absolute cleanliness of all parts is of utmost importance in the operation and servicing of the hydraulic system. Of equal importance is the avoidance of nicks or burrs on any of the working parts.

LUBRICATION

Series 460-560-660

265. The transmission and differential case is also the fluid reservoir for the hydraulic power lift and power steering systems. The fluid capacity is approximately 10 gallons for series 460 tractors and 16 gallons for series

Fig. IH1220—The power steering and hydraulic system filter should be cleaned in kerosene at least every 250 hours of operation.

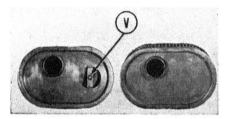

Fig. IH1221—Very early type filter elements were equipped with a by-pass valve (V). A later type filter element shown on right was also used. Refer to Fig. IH1221A for an exploded view of the latest type which is now used.

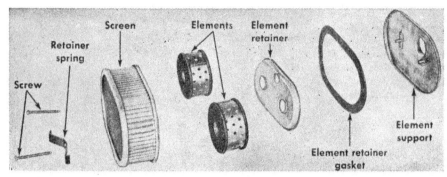

Fig. IH1221A—The latest type filter uses a wire screen and two renewable elements. Use this type filter to replace the two shown in Fig. IH1221.

If either of the early type filters are encountered, it should be discarded and the latest type with the two renewable elements installed.

TESTING

Series 460-560-660

267. The unit construction of the Hydra-Touch system permits removing and overhauling any component of the system without disassembling the others. However, before removing a suspected faulty unit, it is advisable to make a systematic check of the complete system to make certain which unit (or units) are at fault.

NOTE: High pressure (up to 1200-1500 psi) in the system is normal only when one or more of the control valves is in either lift or drop position. When the control valve (or valves) are returned to neutral, the pressure regulator valve is automatically opened and the system operating pressure is returned to a low, by-pass pressure. The low, by-pass pressure is not a factor in the test procedure.

Should improper adjustment, overload or a malfunctioning pressure regulator valve prevent the system from returning to low pressure, continued operation will cause a rapid temperature rise in the hydraulic fluid, damaging "O" rings and seals. Excessively high temperatures will sometimes be indicated by discoloration of the paint on the hydraulic manifold.

Before proceeding with the test, first make certain that the reservoir (transmission case) is filled to the correct level with the proper fluid.

268. Install a pressure gage of sufficient capacity (at least 3000 psi) and a shut-off valve in the pressure line between the pump and the control valves. NOTE: The shut-off valve should be located between the gage and the control valve (refer to Fig. IH1222). With the shut-off valve **open**, start the engine. Move a control valve lever to a raise (or lower) position and notice the pressure as indicated by the gage. At 900-1200 psi the control valve handle should automatically

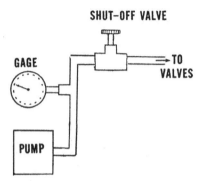

Fig. IH1222—To check the pump and system operating pressures, connect a gage and a shut-off valve in the pressure line as shown and refer to text.

return to the neutral position. With the control valve lever held in the raise (or lower) position, to prevent the lever from moving to neutral, the gage should register 1200-1500 psi.

269. If the pressures as checked in the preceding paragraph 268 are correct, the pump and valves are O.K. and any trouble is located in the lines between the control valves and the cylinder or in the cylinder itself.

If the gage pressure is more than specified, a valve is probably stuck in the closed position.

270. If the pressures were too low when checked as in paragraph 268, proceed as follows: Start engine and with same running, carefully shut the valve off and notice the gage reading which should be at least 1500 psi.

NOTE: The pump may be seriously damaged if the shut-off valve is left in the closed position for more than a few seconds.

If the gage reading is above 1500 psi, with the shut-off valve closed, the pump is O. K. and the trouble may be due to maladjusted and/or leaking relief valves.

If the gage reading remains low with the shut-off valve closed and the filter is known to be O. K., observe the oil. If the oil has a milky appearance, an air leak in the suction line can be suspected. Air leakage at the

---

560 and 660 tractors. International Harvester "Hy-Tran" fluid should be used in the system. The oil should be changed every 1,000 hours of operation or once a year, which ever comes first.

FILTER

Series 460-560-660

266. The working fluid for the hydraulic power lift and power steering systems is also the lubricating oil for the "Torque Amplifier", transmission and differential; therefore, it is very important that the oil be kept clean and free from foreign material. The oil passes through a filter element (Fig. IH1220) before it enters the pump.

The filter should be cleaned in kerosene and renewable elements renewed **at least** every 250 hours of operation. If the tractor remains in operation after the filter is clogged, the filter element may collapse rendering the element useless and possibly damaging the pump (or pumps).

NOTE: Early filters were of the wire screen type and had a by-pass valve (V—Fig. IH1221). Later filters were of the wire screen type but did not have the by-pass valve. The latest type (Fig. IH1221A) incorporates the wire screen and two renewable filter elements.

Fig. IH1224—View of Cessna hydraulic lift and power steering pumps and mounting flange. The longer smooth section of the suction tube (24) should be pressed in the hydraulic lift pump until it bottoms. Although Cessna pumps are shown, Thompson pumps are similar.

pump intake pipe seal can be corrected by removing the pump and mounting bracket and renewing the pipe and seal (24—Fig. IH1224). If, after a reasonable length of time, the oil is still milky, the transmission should be split (detached) from the clutch housing and seal (S—Fig. IH-1164) should be renewed.

TROUBLE-SHOOTING

Series 460-560-660

271. The following trouble shooting chart lists troubles which may be encountered in the operation and servicing of the hydraulic power lift system and should be used after testing as in paragraphs 267 through 270. The procedure for correcting many of the causes of trouble is obvious. For those remedies which are not so obvious, refer to the appropriate subsequent paragraphs.

A. System unable to lift load, gage shows pressure O.K. Could be caused by:
   1. System is overloaded
   2. Damaged hydraulic cylinder
   3. Implement damaged in a manner to restrict free movement
   4. Interference restricting movement of cylinder or implement
   5. Hose couplings not completely coupled

B. System unable to lift load, gage shows little or no pressure. Could be caused by:
   1. Clogged filter
   2. Leaking pump suction line
   3. Faulty pressure regulator valve
   4. Failure of safety valve to close
   5. Leakage past cylinder piston seals
   6. Pump failure

C. System lifts load slowly, gage shows low pressure. Could be caused by:
   1. Clogged filter
   2. Leaking pump suction line
   3. Faulty pressure regulator valve
   4. Failure of safety valve to close
   5. Pressure regulator orifice enlarged or loose in block
   6. Pump failure

D. With all control valves in neutral, gage shows high pressure. Could be caused by:
   1. Pressure regulator piston stuck in its bore
   2. Regulator orifice plugged

E. Operating pressure exceeds 1500 psi. Could be caused by:
   1. Safety valve piston stuck in its bore
   2. Failure of safety valve spring

F. Control valve will not latch in either lift or drop position. Could be caused by:
   1. Broken garter spring
   2. Orifice in rear unlatching piston plugged
   3. Unlatching valve leakage

G. Control valve cannot be readily moved from neutral. Could be caused by:
   1. Control valve gang retaining bolts and nuts too tight, causing valves to bind in valve bodies
   2. Control valve linkage binding
   3. Orifice in rear unlatching piston plugged
   4. Scored control valve and body

H. Control valve unlatches before cylinder movement is completed, gage shows pressure O.K. Could be caused by:
   1. System is overloaded
   2. Damaged hydraulic cylinder
   3. Implement damaged in a manner to restrict free movement
   4. Interference restricting movement of cylinder or implement
   5. Hose couplings not completely coupled

I. Control valve unlatches before cylinder movement is completed, gage shows low pressure. Could be caused by:
   1. Weak unlatching valve spring
   2. Unlatching valve leakage

J. Control valve unlatches from lift position but not from drop position. Could be caused by:
   1. Valve set for single acting cylinder, wherein unlatching valve is inoperative in drop position
   2. Channel between front and rear unlatching valve is plugged

K. Control valve will not center itself in neutral position. Could be caused by:
   1. Control valve gang retaining bolts and nuts too tight, causing valves to bind in valve bodies
   2. Control valve linkage binding
   3. Scored control valve and body
   4. Centering spring weak or broken
   5. Unlatching pistons restricted in movement

L. Control valve will not automatically unlatch from either lift or drop position. Could be caused by:
   1. Clogged filter
   2. Leaking pump suction line
   3. Faulty pressure regulator valve
   4. Plugged channels in control valve which lead to unlatching valve
   5. Leakage past the unlatching pistons
   6. Loose rear unlatching piston retainer
   7. Pump failure

M. Noisy pump operation. Could be caused by:
   1. Clogged filter
   2. Leaking pump suction line
   3. Damaged pump

N. Cylinder will not support load. Could be caused by:
   1. External leakage from cylinder, hoses or connections
   2. Leakage past cylinder piston rings
   3. Internal leaks in control valve

PUMP

Series 460-560-660

Cessna (Fig. IH1225) and Thompson (Fig. IH1226) hydraulic pumps of 12 or 17 gpm capacity are used interchangeably. Both pumps are gear type and are mounted on the manifold (flange). On tractors that are equipped with power steering, the power steering pump is mounted on the rear of the hydraulic pump.

Series 460-560-660

272. **REMOVE AND REINSTALL.** The pump mounting flange (manifold) with the pump or pumps attached can be removed from the left side of the tractor after the hydraulic lift pressure line and, if so equipped, the power steering pressure line are detached from the mounting flange and the flange retaining cap screws are removed.

On tractors so equipped, the power steering pump can be separated from the hydraulic lift pump after removing the four attaching cap screws. Take care to prevent damage to the sealing surfaces when separating power steering pump from the hydraulic lift pump.

On all models the hydraulic lift pump is attached to the mounting flange with four cap screws. Each time the pump is removed from the mounting flange, the sealing rings should be renewed. On tractors so equipped, the cap screws which attach the power steering pump to the hydraulic pump should be torqued to 25 ft.-lbs.

When reinstalling the pump (or pumps) and mounting flange on the tractor, make certain that the longer smooth section of the suction pipe is pressed completely into the hydraulic lift pump. The lip of the seal which is integral with the pipe should face toward the left (pump) side of the tractor. Vary the number and thicknesses of the gaskets which are located between pump mounting flange and clutch housing until the backlash between the pump driven gear and the PTO gear is 0.002-0.022. Gaskets are available in two thicknesses; 0.011-0.013 and 0.021-0.023 for 560 and 660 series tractors, or 0.011-0.019 and 0.016-0.024 for 460 series tractors.

273. OVERHAUL CESSNA. To overhaul the removed Cessna pump, first remove the power steering pump, or cover (14—Fig. IH1225), from rear of pump and nut, drive gear and key (7) from front. The remainder of the disassembly procedure will be evident.

The bushings in the cover and body are not available separately. The pumping gears, shafts and snap rings are available only as a complete set (8).

12 GPM Pump

O.D. of shafts at bushings..0.810 min.
I.D. of bushings in body
and cover ....0.816 max.
Thickness of gears........0.572 min.
I.D. of gear pockets in body 2.002 max.

17 GPM Pump

O.D. of shafts at bushings..0.810 min.
I.D. of bushings in body
and cover ....0.816 max.
Thickness of gears........0.813 min.
I.D. of gear pockets in body 2.002 max.

When reassembling, use new diaphragms (10 and 18), phenolic gasket (17), back-up gasket (16), diaphragm seal (15), ball (19), spring (20) and all oil seals. With open part of diaphragm seal (15) towards cover (2) work same into grooves of cover using a dull tool. Press the phenolic gasket (16) and the back-up gasket (17) into

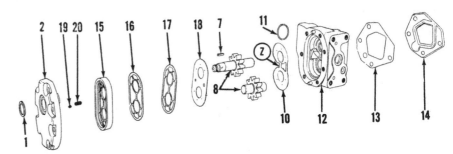

Fig. IH1225—Exploded view of the Cessna hydraulic system pump. Pumps of 12 and 17 gpm capacities are available.

1. Seal	10. Diaphragm	14. Cover	18. Diaphragm
2. Cover	11. "O" ring	15. Diaphragm seal	19. Ball
7. Key	12. Body	16. Back-up gasket	20. Spring
8. Pumping gears and shafts	13. Gasket	17. Phenolic gasket	

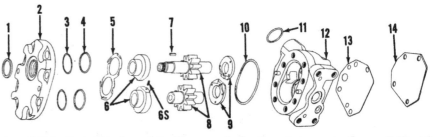

Fig. IH1226—Exploded view of the Thompson hydraulic system pump. Pumps of 12 and 17 gpm capacities are available.

1. Seal	4. "O" rings (2 used)	8. Pumping gears and shafts	11. "O" ring
2. Cover	5. Spring	9. Bearings	12. Body
3. Back-up washers (2 used)	6. Pressure bearings	10. "O" ring	13. Gasket
	7. Key		14. Cover

the relief in the diaphragm seal (15). Install the check ball (19) and spring (20) in cover; then, install the diaphragm (18) with bronze face toward gears. NOTE: The diaphragm (18) must fit inside the raised rim of the diaphragm seal (15). Dip gear and shaft assemblies (8) in oil and install them in cover. Position the diaphragm (10) in the body (12) with the bronze side toward gears and the cut-out (Z) toward inlet side of pump body. Install the pump body over the gears and shafts. Check the pump rotation. Pump should have a slight amount of drag but should rotate evenly.

274. OVERHAUL THOMPSON. To overhaul the removed Thompson pump, first remove the power steering pump, or cover (14—Fig. IH1226), from rear of pump and nut and drive gear key (7) from front. Remove cap screws retaining cover to pump body, bump pump drive shaft on a wood block to loosen cover from pump body and remove cover. Remove seal rings (4), fiber washers (3), spring (5) and ring gasket (10). Press drive shaft seal (1) out of pump cover. Tap drive shaft on a wood block to loosen bearings (6) from pump body, then remove the bearings.

Remove gears (8). Tap body on wood block to remove bearings (9).

Identify the bearings so they can be installed in their original position.

Clean all metal parts in a suitable solvent and dry them with compressed air. If seal contacting surfaces on drive gear shaft are not perfectly smooth, polish them with fine crocus cloth and rewash the drive gear and shaft.

Dimensions for both pumps are: Shaft O.D., 0.812 (min.); bearing I.D., 0.816 (max.); gear pocket I.D., 1.772 (max.); gear width (12 gpm), 0.7765 (min.); gear width (17 gpm), 1.072 (min.)

When reassembling, lubricate all parts with clean oil and use new gaskets and seals.

Install bearings (9) in their original position with milled slot on pressure side. Install gears and shafts (8) in the pump body. Install bearings (6), seal ring (10) and the spring (5). Install seal (1) in cover so that lip of seal faces center of pump. Install back up washer (3) and seal ring (4) in cover. Install pump cover carefully to avoid damaging the seal. Install cover cap screws and tighten them to a torque of 25 ft.-lbs. Install the drive shaft Woodruff key and drive gear. Check the pump rotation. Pump should have a slight amount of drag but should rotate evenly.

Fig. IH1230—Right side view of two hydraulic control valves and a follow-up ("Tel-A-Depth") control valve installed on a 460 International tractor.

Fig. IH1231 — Left side view of two hydraulic control valves and a "Tel-A-Depth" control valve installed on a 460 International tractor.

Fig. IH1232 — Right side view of two hydraulic control valves and a "Tel-A-Depth" control valve installed on a 560 Farmall tractor.

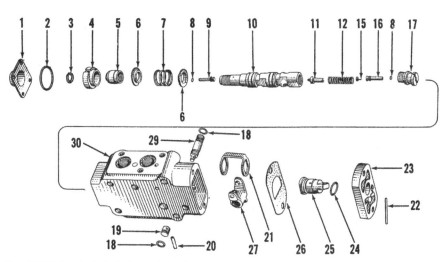

Fig. IH1233—Exploded view of the hydraulic system ("Hydro-Touch") control valve. The garter spring (3) may be one piece or two pieces.

1. Valve cap
2. Seal ring
3. Garter spring
4. Garter spring sleeve
5. Front unlatching piston retainer
6. Control valve centering spring retainers
7. Control valve centering spring
8. Seal rings
9. Front unlatching piston
10. Control valve spool
11. Unlatching valve
12. Unlatching valve spring
15. Control valve orifice plug and screen
16. Rear unlatching piston
17. Rear unlatching piston retainer
18. Seal rings
19. Bushing
20. Roll pin
21. Guide
22. Roll pin
23. Body cover
24. Seal ring
25. Indexing bushing
26. Gasket
27. Yoke
29. Control valve lever shaft
30. Control valve body

CONTROL VALVES

Series 460-560-660

Tractors may be equipped with either a single, double or triple valve system. All valves are identical on multiple valve systems.

275. REMOVE AND REINSTALL. To remove the control valve (or valves), remove the interfering sheet metal and, on Farmall tractors so equipped, index mark the follow-up control lever and shaft. Remove the roll pin and pull lever from the shaft. Detach the follow-up control bracket from the control valve left support. On all models, disconnect the hydraulic lines from the left support, regulator valve block, and control valves. Remove the cap screws which attach the control valve left support to the transmission cover and the steering shaft and dash support. Unscrew the four cap screws which extend through the valves and into the left support and remove the left support.

Detach the control linkage from the control valve levers and remove the control valves.

On International tractors, the regulator and safety valve block is mounted on the same cap screws as the control valves.

When reinstalling the valves, renew all "O" rings and seals and reverse the removal procedure. On models so equipped, the follow-up cable may need to be adjusted as in paragraph 280.

276. OVERHAUL. Thoroughly clean the removed control valve unit in a suitable solvent, refer to Fig. IH1233 and proceed as follows:

Remove body cover (23) and gasket (26). Drift out roll pin (22) and remove indexing bushing (25). Remove roll pin (20) and withdraw lever shaft (29), bushing (19) and yoke (27). NOTE: Depending on the application, yoke (27) and lever shaft (29) may be turned 180 degrees from that shown in Fig. IH1233. Lift out valve guide (21). Remove cap (1), garter spring sleeve (4) and garter spring (3). Push the control valve spool assembly from the valve body. Clamp the spool in a soft jawed vise and remove the rear retainer (17); then, withdraw the unlatching valve (11) and its spring (12). Remove the unlatching piston (16) from the rear retainer. Unscrew the orifice plug (15) from piston (16). Turn the spool over in the vise and remove the front retainer (5), retainers (6), centering spring (7) and front unlatching piston (9).

The unlatching valve spring (12) should have a free length of 1⅝ inches and should test 12 lbs. at ⅞-inch. The garter spring (3) should be renewed if outer diameter shows excessive wear. Note: Garter springs in later control valves are of two-piece construction but are interchangeable with early springs. Renew the matched spool and body units if clearance is excessive or if either part shows evidence of scoring or galling. Thoroughly clean channels and the small bores in the valve spool (10). Inspect the unlatching valve (11) and its seat in the control valve spool for damage.

Clean the orifice plug and screen (15) with compressed air and make certain that bore in rear piston (16) as well as the passages in body (30) are open and clean.

When reassembling, dip all parts in clean hydraulic fluid, then use new "O" ring seals and gaskets and reverse the disassembly procedure. Special bullet tool No. ED-3396 will facilitate installation of the garter spring (3) and spring sleeve (4). When installing guide (21), make certain that strap of guide is toward same side of body as shown in Fig. IH1233. Install yoke (27) with serrated end of yoke bore toward same side of body as bushing (19). Insert shaft (29) with pin hole up; then, install bushing (19) and roll pin (20). Install body cover (23) so that indexing bushing (25) engages guide (21). Install the small lever on outer serrations of shaft (29) in its original position. Note: On multiple valve systems, the relative position of the lever with respect to shaft (29) must be the same on all valves.

When installing the control valve, tighten the retaining bolts to a torque of 25 ft.-lbs. Over-tightening will result in distorted body and binding valve spool.

REGULATOR AND SAFETY VALVE BLOCK

Series 460-560-660

277. REMOVE AND REINSTALL. The procedure for removing the regulator and safety valve block from 460 and 560 Farmall tractors will be self-evident.

On 460, 560 and 660 International tractors, follow the procedure outlined in paragraph 275 to remove the valve block.

278. OVERHAUL. Thoroughly clean the removed valve unit in a suitable solvent, refer to Fig. IH1234 and proceed as follows:

Remove two diagonally opposite cap screws retaining the cover (1) to the body and insert in their place 2-inch long cap screws and tighten them finger tight. Remove the two remaining short cap screws; then, relieve the pressure of the safety valve spring gradually by alternately unscrewing the 2-inch long screws. Remove cover (1) and gasket (3). Withdraw the safety valve spring (11), spring retainer (12) and safety valve (14). Remove the regulator valve piston (5), unscrew the valve seat (6) and remove the ball (7), rider (8) and spring (9). Unscrew and remove the orifice plug and screen (10).

Inspect ball valve (7) and valve seat (6) for damaged seating surfaces. Piston (5) and safety valve (14) must be free of nicks or burrs and must not bind in the block bores. Clean the orifice plug and screen (10) with compressed air and make certain that bores in block (15) are open and clean.

When reassembling, dip all parts in clean hydraulic fluid, use new "O"

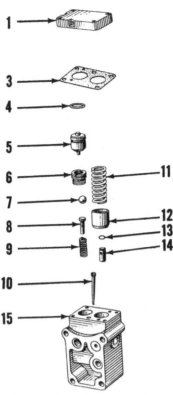

Fig. IH1234—Exploded view of the regulator and safety valve block used on models with "Hydra-Touch" system.

1. Cover	9. Ball rider spring
3. Gasket	10. Safety valve orifice
4. Seal ring	screen and plug
5. Regulator valve	11. Safety valve spring
piston	12. Spring retainer
6. Regulator valve	13. Snap ring
seat	14. Safety valve piston
7. Steel ball	15. Valve housing
8. Ball rider	

ring seals and gaskets and reverse the disassembly procedure. Be sure that all openings in block, gasket (3) and cover (1) are aligned.

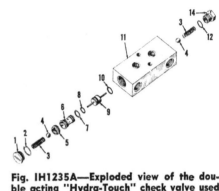

Fig. IH1235 — Exploded view of the single acting "Hydra-Touch" system check valve used on some models.

1. Plug	6. Ball
2. "O" ring	7. Check valve spring
3. Actuator piston	8. "O" ring
4. "O" ring	9. Plug
5. Block	

Fig. IH1235A—Exploded view of the double acting "Hydra-Touch" check valve used on some models.

1. Plug	8. "O" ring
2. "O" ring	9. Piston
3. Spring	10. "O" ring
4. Check ball	11. Valve block
5. Retainer nut	12. "O" ring
6. Retainer	14. Plug
7. "O" ring	

CHECK VALVE

Models So Equipped

279. The procedure for removing, overhauling and/or cleaning the check valve is evident after an examination of the unit and reference to Fig. IH1235 or IH1235A.

TEL-A-DEPTH SYSTEM

A Tel-A-Depth hydraulic system is available on series 460, 560 and 660 tractors. The system provides a means whereby the implement returns to the previously determined working depth which has been selected on the control quadrant. An adjustable stop on the quadrant allows the operator to select any desired working depth and to return to the same depth by moving control lever against stop. The stop can also be by-passed should the operator desire.

When a tractor is equipped with the Tel-A-Depth system, the Tel-A-Depth valve replaces the Hydra-Touch control valve used for rear implements. Two types of Tel-A-Depth valves are used as shown in Figs. IH1244 and IH1244A.

Refer to Fig. IH1240, IH1240A or IH1240B for schematic view of the Tel-A-Depth system.

SYSTEM ADJUSTMENT

Early (Cable Linkage) Type

280. Check hydraulic fluid reservoir (transmission case) and fill to proper level, if necessary. Start engine and cycle system several times; then, place hand control lever in the forward position. The system should go off pressure and the distance between center of cylinder rod pin and face of cylinder should be $1\frac{3}{4}$-$1\frac{7}{8}$ inches as shown in Fig. IH1241. Now move the hand control lever to the rearward position. The system should again go

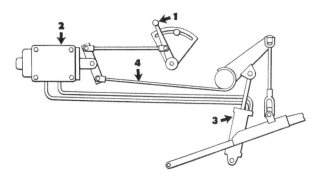

Fig. IH1240 — Schematic drawing of a typical early (cable linkage) type "Tel-A-Depth" system. Lever (1) opens valve (2) which actuates cylinder (3) to raise or lower implement. The follow-up cable (4) attached to rockshaft closes valve (2).

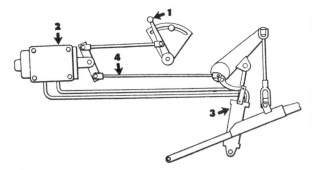

Fig. IH1240A — Schematic drawing of a "Tel-A-Depth" system typical of that used on late 460 Farmall, 560 Farmall and International and 660 International tractors. Earlier models may be changed to this later type. Lever (1) opens the valve (2) and movement of the rockshaft closes it again.

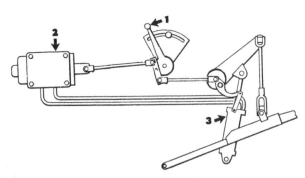

Fig. IH1240B — Schematic drawing of a "Tel-A-Depth" system typical of that used on late 460 International tractors. Earlier models may be changed to this later type.

Fig. IH1241—With hand control lever in the forward position, system should be off pressure and distance between face of cylinder and centerline of cylinder rod pin should be 1¾-1⅞ inches.

1. Compensating lever
2. Ball joint
3. Compensating rod
4. Scale
5. Piston rod
6. Gear box

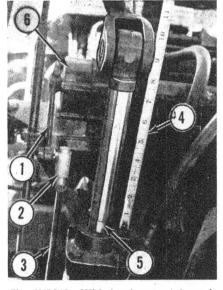

Fig. IH1242—With hand control lever in the rearward position, system should be off pressure and distance between face of cylinder and centerline of cylinder rod pin should be 9⅜-9½ inches. Refer to Fig. IH1241 for legend.

off pressure and the distance between center of cylinder rod pin and face of cylinder should be 9⅜-9½ inches as shown in Fig. IH1242.

If above conditions are not met, adjust the system as follows: Start tractor engine and place hand control lever in its forward position on the quadrant. Disconnect the ball joint from the valve follow-up lever and fully retract (collapse) hitch cylinder by moving the follow-up lever rearward. Shut off engine and be sure cylinder remains retracted. Loosen lock nut under ball joint on compensating rod (3—Fig. IH1241); then, disconnect ball joint (2) from compensating lever (1). Move compensating lever downward to be sure that rockshaft drive unit slip clutch is engaged. If slip clutch is not engaged, a click will be heard as compensating lever is moved downward and the clutch engages. NOTE: Slip clutch should engage every 90 degrees. If clutch does not engage, on the downward movement of the compensating lever, move lever in the opposite (upward) direction until it does engage. Pull upward on compensating lever and pull follow-up cable into conduit until approximately ⅜-inch of cable is left exposed between end of conduit and back of lock nut at valve end of cable. Hold the compensating lever in this position to maintain the ⅜-inch dimension, then, adjust length of compensating rod so that ball joint can be engaged on ball of compensating lever without binding. In some cases, it may be necessary to reposition the compensating lever on the splined shaft which extends from the rockshaft drive unit (6—Fig. IH1241). When this occurs, mark the position of lever on shaft prior to removal of lever and re-establish the previously mentioned ⅜-inch dimension after lever is reinstalled. Tighten ball joint lock nut on compensating rod and, if necessary, the cap screw on compensating lever; then, reconnect ball joint to follow-up lever on valve.

Start tractor engine, cycle system several times and place hand control lever in the forward position. If system remains on pressure with hand control forward and cylinder retracted (collapsed), loosen conduit pivot clip and move the conduit back and forth until an off pressure position is found and the distance between center of cylinder pin and cylinder face is 1¾-1⅞ inches. Tighten pivot clip. Cycle system several times and recheck cylinder position with hand lever in forward position. If additional adjustment is needed to maintain the 1¾-1⅞ inches cylinder measurement, the quadrant limit stop can be adjusted until proper cylinder measurement is obtained.

NOTE: The ⅜-inch exposed cable measurement may not be maintained after making above adjustments.

With engine running, place hand control in rearward position, check to see that system is off pressure and that the distance between center of cylinder pin and face of cylinder is 9⅜-9½ inches. Adjust the quadrant stop to obtain the proper cylinder measurement.

Latest (Rod Linkage) Type

281. Check the hydraulic fluid in the reservoir (transmission case) and fill to the proper level, if necessary. Start engine and cycle system several times; then, place hand control lever in the forward position. The system should go off pressure and the distance between center of cylinder rod pin and face of cylinder should be 1¾-1⅞ inches. (Refer to Fig. IH1241 which shows the same measurement of the early type.) Now move the hand control lever to the rearward position. The system should again go off pressure and the distance between center of cylinder rod pin and face of cylinder should be 9⅜-9½ inches.

On all tractors except 460 International, if the system remains on pressure when the distance between the center of the cylinder rod pin and the face of the cylinder is between 1¾-1⅞ inches, **lengthen** the input rod (4—Fig. IH1240A and IH1249A) until the system goes off pressure with 1¾-1⅞ inches dimension. If the cylinder does not fully retract, input rod should be **shortened**.

On International 460 tractors, if the system remains on pressure when the distance between the cylinder face and the center of the cylinder rod pin is 1¾-1⅞ inches, the front actuating rod (AR—Fig. IH1251) should be **shortened.** If the system goes off pressure before the distance is 1⅞ inches, the front actuating rod should be **lengthened.**

On all models, move the "Tel-A-Depth" control handle completely to the rear of the quadrant. If the system stays on pressure when the distance between the center of the cylinder rod pin and the face of the cylinder is between 9⅜-9½ inches or; if the center of the cylinder rod pin is not 9⅜-9½ inches from the face of the cylinder, adjust the quadrant stop.

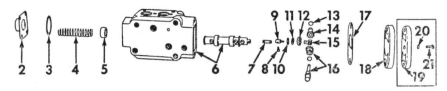

Fig. IH1244—Exploded view of a typical differential gear type "Tel-A-Depth" control valve. Lockout cover assembly (items 19, 20 and 21) may be used on some applications.

2. Cap	8. Pin	13. "O" ring	17. Gasket
3. "O" ring	9. Yoke link	14. Control sleeve and gear	18. Cover
4. Centering spring	10. Washer		19. Lockout cover
5. Spring cup	11. Shim (light and heavy)	15. Eyebolt shaft	20. "O" ring
6. Body and spool		16. Follow-up shaft and gear	21. Lockout screw
7. Spool link	12. Idler gear		

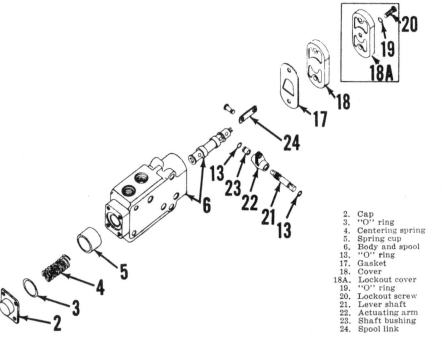

Fig. IH1244A—Exploded view of the "Tel-A-Depth" valve used when a walking beam type of control linkage, such as that used on International 460, is used. Lockout cover assembly (items 18A, 19 and 20) may be used on some applications.

2. Cap
3. "O" ring
4. Centering spring
5. Spring cup
6. Body and spool
13. "O" ring
17. Gasket
18. Cover
18A. Lockout cover
19. "O" ring
20. Lockout screw
21. Lever shaft
22. Actuating arm
23. Shaft bushing
24. Spool link

TEL-A-DEPTH CONTROL VALVE

Series 460-560-660

282. **R&R AND OVERHAUL.** Disconnect control rods from control valve input and follow-up levers. Remove the nuts from the through bolts which hold ganged valves to valve support and if necessary, remove the outer "Hydra-Touch" control valve. Withdraw through bolts from right side of tractor until Tel-A-Depth valve is free; then, pull valve downward and remove manifold from top side of valve. Use caution during this operation not to deform manifold.

With valve removed, refer to Fig. IH1244 and IH1244A. Disassembly is as follows: Remove valve cover (18—Fig. IH1244 or IH1244A). Valve spool can be removed for cleaning and/or inspection by removing pin which retains spool to yoke. If spool and/or valve body are damaged, renew the

Fig. IH1245 — View showing shafts and gears removed from differential type "Tel-A-Depth" valve.

valve assembly as spool and body are mated parts. On valves so equipped, remove gears and shafts by driving roll pins from hubs of control lever and follow-up shaft gears. Unscrew yoke from link and separate yoke, idler gear and link. If used, save the shims located between yoke washer and idler gear. Refer to Fig. IH1245. On late 460 International tractors, refer to Fig. IH1244A, remove levers and shafts.

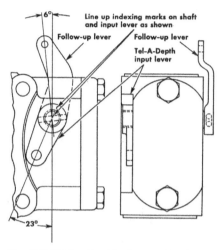

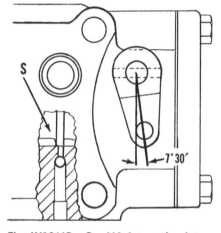

Fig. IH1246 — On 460 Farmall tractors equipped with the early (cable) type linkage, use this drawing as a guide when reassembling the "Tel-A-Depth" valve.

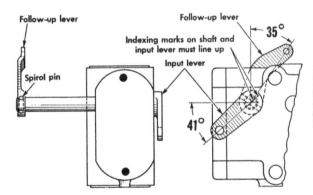

Fig. IH1246A—On 460 International tractors equipped with the early (cable) type linkage, use this drawing when reassembling the "Tel-A-Depth" valve.

Fig. IH1246D—On 460 International tractors equipped with the late (rod) type linkage, use this drawing when reassembling the "Tel-A-Depth" valve. Spool (S) is centered.

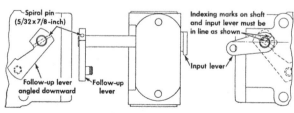

Fig. IH1246B — On 560 Farmall tractors equipped with the early (cable) linkage, use this drawing when reassembling the "Tel-A-Depth" valve.

Fig. IH1246C — On 460 Farmall tractors equipped with the late (rod) type linkage, use this drawing as a guide when reassembling the "Tel-A-Depth" valve.

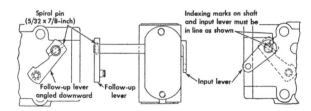

Fig. IH1246E — On 560 Farmall tractors equipped with the late (rod) type linkage, use this drawing when reassembling the "Tel-A-Depth" valve.

NOTE: If valve is the earliest type, without markings, pay close attention to the position of gear yoke, input lever and follow-up lever. Place correlation marks on shafts and levers prior to removing the levers.

On later valves, the idler gear has two index marks 180 degrees apart on back face of gear and the input and follow-up gears have one index mark on the outside diameters. In addition, the input and follow-up levers and their shafts have index marks to insure correct reassembly. Refer to Fig. IH1246, 1246A, 1246B, 1246C or 1246E.

On all models, centering spring (4—Fig. IH1244 or IH1244A) and spring cup (5) can be removed after moving cap (2).

Inspect all parts for damage and/or wear. If input shaft, input gear, follow-up shaft, follow-up gear or idler gear are to be renewed, it will be necessary to renew complete gear assembly as the above parts are not catalogued separately.

Reassembly is the reverse of disassembly, however, it is recommended that new "O" rings be used. Refer to Fig. IH1246, IH1246A, IH1246B, IH-IH1246C, IH1246D or IH1246E.

On models with drive gears, adjust yoke until idler gear is snug yet will turn freely using shims (11—Fig. IH-1244), if necessary. On the early unmarked valves be sure parts are reassembled in the same position as they were originally. Axis of yoke and gears must be parallel to cover surface. On later marked valves, mate all index marks. Refer to Fig. IH1247 for an assembled view of the differential gear type valve.

On all models, it may be necessary to readjust system as outlined in paragraph 280 or 281.

ROCKSHAFT DRIVE UNIT

Early (Cable Linkage) Type

283. **R&R AND OVERHAUL.** Remove conduit shield. Disconnect ball joint from valve follow-up lever and remove pivot clip retaining cap screw. Remove conduit from retaining clips located along tractor rear frame. Dis-

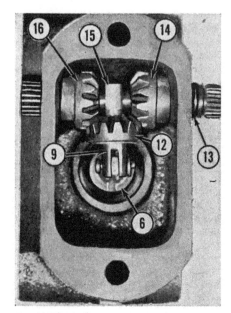

Fig. IH1247—View of "Tel-A-Depth" valve typical of all except late 460 International (refer to Fig. IH1244A). The shaft on the right side is partially removed to show "O" ring (13).

6. Spool	15. Eyebolt (idler)
9. Yoke	shaft
12. Idler gear	16. Follow-up shaft
13. "O" ring	gear
14. Control lever gear	

connect compensating rod from compensating lever and remove lever. Unbolt rockshaft drive unit from rockshaft drive side plate and pull unit rearward. Remove ball joint and rubber boot from valve end of cable assembly. Loosen conduit retaining nut from drive unit and remove cable and conduit by turning cable counterclockwise. Pull cable from conduit. Remove cover from drive unit, which in some cases will require removal of staking. Remove parts from drive unit as shown in Fig. IH1248.

Clean all parts and inspect same for excessive wear and/or damage.

NOTE: Early units had gears made of powdered metal while later gears are steel. If necessary to renew the early type gears, order IH service package number 373 623 R91. Other parts are available as service items.

With cable and conduit cleaned, lubricate cable with Lubriplate (105-V) or equivalent, and install cable in conduit. Temporarily install ball joint on threaded end of cable and attach a low reading (1-10 lbs.) spring scale to same. Measure effort required to move cable through conduit and if effort exceeds four pounds, renew cable and conduit assembly.

Lubricate all drive unit parts with Lubriplate (105-V) or equivalent, and proceed as follows: Place drive wheel in housing and position same so cable

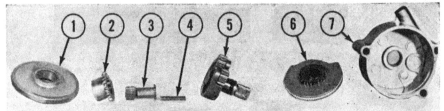

Fig. IH1248—View showing the component parts of the rockshaft drive unit (gear box) and their relative position. Later tractors which are equipped with the rod linkage do not use this drive unit.

1. Cover	3. Coupling	5. Planet gear assembly	6. Cable drive wheel
2. Input gear	4. Spring		7. Housing

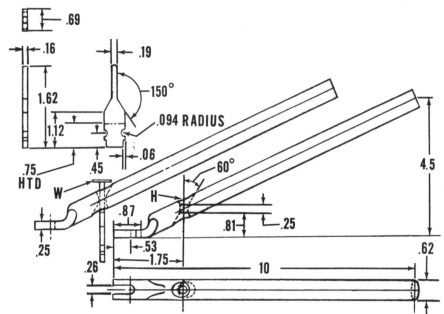

Fig. IH1249—View of special tool used to facilitate removal of rockshaft slip clutch. Tool can be made locally. "H" is hole of 0.250 diameter. "HTD" means harden this distance. "W" is washer.

Fig. IH1249A—View of the left side of a 460 Farmall tractor showing the late rod linkage installed. The input rod is shown at (4). Farmall 560 and International 560 and 660 tractors so equipped are similar.

anchor hole aligns with cable lead-in hole in housing. Hold wheel in position, install cable and turn cable clockwise until it bottoms. Rotate

Fig. IH1249B—View of the compensating arm and rockshaft pivot plate installed on a late 460 Farmall tractor. Farmall 560 and International 560 and 660 tractors so equipped are similar.

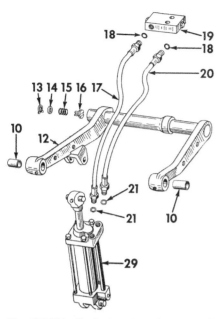

Fig. IH1251—View of the left side of a 460 International tractor showing the late rod linkage installed. The front actuating rod is shown at (AR).

Fig. IH1250—Typical rockshaft and "Tel-A-Depth" cylinder showing slip clutch exploded from rockshaft.

10. Bushing	16. Coupling
12. Rockshaft	18. Ring seal
13. Snap ring	19. Check valve
14. Retainer	21. Ring seal
15. Spring	29. Cylinder

wheel and push cable into housing until conduit is positioned; then, tighten conduit retaining nut. Balance of reassembly is the reverse of disassembly. The planet gear carrier may be installed in any position with respect to the drive wheel.

After installing compensating lever and prior to installing the compensating rod, check operation of rockshaft slip clutch as follows: Pull up on compensating lever until ball joint butts against conduit. Attach a spring scale to lever and while keeping scale at 90 degrees to compensating lever, measure the force required to cause clutch to slip. This force should be 38 to 50 pounds. If not within these limits, refer to paragraph 284.

Connect compensating rod and lever and adjust system as outlined in paragraph 280.

ROCKSHAFT SLIP CLUTCH

Early (Cable Linkage) Type

284. **R&R AND OVERHAUL.** To remove the rockshaft slip clutch, disconnect the compensating rod from compensating lever, loosen the conduit retaining nut, then, unbolt the drive unit from the rockshaft side plate and swing unit away from rockshaft. Use a tool such as that shown in Fig. IH-1249 and compress clutch spring and ring retainer into rockshaft. Remove the retaining snap ring, slowly release

tool and remove ring retainer, spring and coupling. See Fig. IH1250.

Clean all parts including the coupling insert which is pressed into rockshaft. Inspect all parts for excessive

wear and/or damage. Pay particular attention to mating surfaces of clutch coupling and insert and remove any burrs which may be present.

Apply a light coat of light oil to clutch faces and inside bore of rockshaft; then, reassemble unit by reversing the disassembly procedure. Check operation of clutch after assembly as follows: Use a torque wrench fitted with a screw driver adaptor and check the torque required to cause clutch to slip. Clutch should slip in either direction of rotation between 80-100 in.-lbs. Turn clutch coupling 180 degrees and repeat operation. If clutch does not meet the above specifications, renew assembly.

HYDRAULIC LIFT SYSTEM (DRAFT AND POSITION CONTROL)

Series 606-2606

290. The series 606 and 2606 hydraulic lift unit, which provides both draft and position control, and referred to as "Draft and Position Control", also serves as a top cover for the differential and final drive portions of the tractor rear center frame. Contained within the hydraulic lift housing are the rockshaft, work cylinder, control valves and the linkage for operation of the components. See Fig. IH1257.

Pressurized oil for the operation of the hydraulic lift unit is supplied by a transmission driven pump (13—Fig. IH1258). Pumps for hydraulic lift unit may be either 4½ or 7 gpm capacity, depending upon usage of tractor. In addition, tractors with power steering and auxiliary circuits are equipped with either 12 or 17 gpm pumps (10) mounted on the same pump mounting flange with the hydraulic lift pump. Pumps operate at a pressure of 1550-1600 psi, controlled by the system pressure relief valves. A hydraulic fluid filter is mounted in right side of the clutch housing as shown in Fig. IH1255.

The hydraulic lift system draws its oil supply from the transmission and differential housing and thus, the tractor rear center frame and hydraulic lift system share a common reservoir.

LUBRICATION AND BLEEDING

Series 606-2606

291. To drain and refill the reser-

voir, and bleed the hydraulic system, proceed as follows: Remove drain plug from rear center frame, torque-amplifier housing and right rear bottom of hydraulic lift housing. With switch, or fuel shut-off in the off position, crank engine briefly to clear pump and connecting lines.

Remove filler plug from transmission top cover and level plug from right side of rear center frame. Fill rear center frame to level plug opening with IH Hy-Tran Fluid. Start engine and with filler plug out, cycle lift system, and remote cylinders if so equipped, about ten or twelve times, then place position control lever and remote control levers in the forward position. Cycle steering system from one extreme position to the other, then place wheels in straight ahead position. Recheck fluid level and add as necessary to bring to level plug opening. Install and tighten level plug and filler plug.

FILTER

Series 606-2606

292. The working fluid for the hydraulic power lift and power steering systems is also the lubricating oil for the "Torque Amplifier", transmission and differential; therefore, it is very important that the oil be kept clean and free from foreign material. The oil passes through a filter element (Fig. IH1255) before it enters the pump. Also see Fig. IH1256.

The filter should be cleaned in kerosene and renewable elements renewed **at least** every 250 hours of operation.

Fig. IH1255—The hydraulic and power steering system filter is located in right hand side of clutch housing as shown.

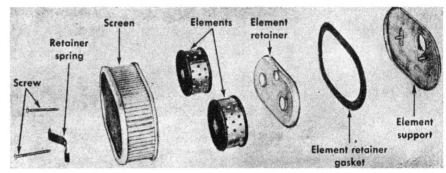

Fig. IH1256—Exploded view of hydraulic and power steering system filter.

If the tractor remains in operation after the filter is clogged, the filter element may collapse rendering the element useless and possibly damaging the pump (or pumps).

TROUBLE SHOOTING

Series 606-2606

293. The following trouble-shooting chart lists troubles which may be encountered in the operation and servicing of the hydraulic lift system. The procedure for correcting many of the causes of trouble is obvious. For those remedies which are not so obvious, refer to the appropriate subsequent paragraphs.

A. Hitch will not raise. Could be caused by:
1. Unloading valve orifice plugged.
2. Unloading valve piston sticking.
3. Flow control valve spring broken, piston sticking or check ball stuck in orifice.
4. Unloading valve ball not seating.
5. Main relief valve spring broken or valve is leaking.
6. Cylinder safety (cushion) valve faulty.
7. Auxiliary valve cover "O" ring damaged.
8. Linkage disconnected from control lever or valve.

B. Hitch lifts load too slow. Could be caused by:
1. Flow control valve piston sticking or faulty valve spring.
2. Flow control valve piston stop broken.
3. Unloading valve ball seat leaking.
4. Faulty main relief valve.
5. Cylinder safety (cushion) valve leaking.
6. Scored lift cylinder or piston "O" ring damaged.
7. Excessive load.

C. Hitch will not lower. Could be caused by:
1. Control valve spool sticking or "O" ring damaged.
2. Drop poppet valve sticking or "O" ring damaged.

D. Hitch lowers too slow. Could be caused by:
1. Drop control valve spool sticking.
2. Damaged "O" ring on drop poppet valve.
3. Linkage out of adjustment.

E. Hitch lowers too fast. Could be caused by:
1. Faulty drop control valve piston.

F. Hitch lowers too fast in slow action position. Could be caused by:
1. Drop control linkage out of adjustment.

G. Hitch raises and lowers but will not maintain position (hiccups). Could be caused by:
1. Check valve in main control valve leaking.
2. Cylinder safety (cushion) valve leaking.
3. Cylinder scored or piston "O" ring damaged.
4. Drop poppet valve sticking.
5. Check valve actuating rod adjusting screw improperly adjusted.

H. System stays on high pressure. Could be caused by:
1. Broken, disconnected or improperly adjusted linkage.
2. Control valve spools faulty.
3. Auxiliary valve not in neutral.
4. Restraining chains adjusted too short.
5. Core plug in control valve body missing or leaking badly.

I. Malfunction of position control. Could be caused by:
1. Control valve spool or linkage binding.

2. Control lever quadrant not mounted correctly.
3. Incorrect valve link-to-spool adjustment.

J. Response time too slow. Could be caused by:
1. Unloading valve piston sticking.
2. Unloading valve orifice plugged (partially).

K. Hitch has too much depth variations (over-travels). Could be caused by:
1. Torsion bar bearing not lubricated.
2. Flow control valve check ball missing.
3. Unloading valve orifice partially plugged.

L. Inadequate depth control during deep operation. Could be caused by:
1. Foreign material between torsion bar bellcrank and its stops.
2. Incorrect adjustment of hitch lift link.
3. Interference between top link and rear frame.
4. Control lever quadrant mounted in wrong position.
5. Top link pin excessively worn.

M. Insufficient transport clearance of mounted implement. Could be caused by:
1. Incorrect valve link-to-control valve spool adjustment.
2. Safety shut-off improperly adjusted.
3. Incorrect adjustment of hitch lift link.

N. Auxiliary circuit will not lift load or lifts load slowly. Could be caused by:
1. Faulty main relief valve.
2. Excessive leakage past valve spool.
3. Faulty auxiliary valve check ball.
4. Faulty cylinder relief valve (industrial valve),

O. Auxiliary valves does not automatically unlatch. Could be caused by:
  1. Incorrect unlatching adjustment.
  2. Faulty unlatching piston.
  3. Faulty detent sleeve.
  4. Spool sticking due to improperly tightened mounting bolts.

P. Auxiliary valve unlatches prematurely. Could be caused by:
  1. Incorrect unlatching adjustment.
  2. Broken unlatching spring.
  3. Faulty detent sleeve.

Q. Load drops slightly when auxiliary valve is put in lift position. Could be caused by:
  1. Valve check ball leaking or not seating.

R. Auxiliary valve will not hold load in position. Could be caused by:
  1. Excessive leakage past valve spool.
  2. Faulty cylinder or piston.
  3. Faulty cylinder relief valve (industrial valve).

S. Fluid leaking from detent breather. Could be caused by:
  1. Faulty unlatching piston "O" ring.
  2. "O" ring at detent end of valve spool leaking.

SYSTEM OPERATING PRESSURE AND RELIEF VALVE

Series 606-2606

294. The overall system operating pressure can be determined with the lift system pressure relief valve (in lift housing) and the auxiliary system relief valve (in transfer block) installed in tractor but as both relief valves are involved in the system operation it is difficult to determine which relief valve is involved when system is malfunctioning.

If a flow meter, such as the International Harvester Flo-Rater, or its equivalent, is available, an indication of which valve is malfunctioning can be determined by the amount of drop in gpm. This gpm drop will correspond to the size of pumps which the tractor is equipped with. For example, if the gpm drop is something between 4½ and 7 gpm it would indicate that the lift system relief valve is opening too soon. If the gpm drop is something between 9 and 14 gpm, it would indicate the auxiliary circuit relief valve is opening too soon.

If the overall system operating pressure test results in a low reading, it is recommended that both valves be removed and bench tested.

The overall system operating pressure can be checked as follows: Use a gage capable of registering at least 2000 psi and install gage in series with a shut-off valve in a line to which has been attached the male half of a quick coupler. Gage must be in the line between shut-off valve and male half of quick coupler. Install test assembly male end in quick coupler of tractor and place open end in reservoir filler hole. Be sure to fasten open end securely in filler hole. Start engine and operate until hydraulic fluid is warmed to operating temperature; then, with engine operating at high idle speed, move auxiliary lever to the position that will direct fluid to test gage. Manually hold control valve lever in this position and close the shut-off valve only long enough to observe the pressure reading on the gage. Pressure reading should be 1550-1600 psi.

NOTE: Before removing test fixture, test auxiliary control valve unlatching pressure as follows: Operate engine at low idle speed, move auxiliary control valve lever to position that will direct fluid to test gage, then slowly close the shut-off valve and observe the pressure at which the valve control lever returns to neutral. This pressure should be not less than 1000 psi nor more than 1250 psi.

If the overall system pressure is not as specified, refer to paragraph 295 for information concerning the lift system relief valve and to paragraph 314 for information concerning the pilot relief (auxiliary circuit) valve.

For information concerning unlatching mechanism of auxiliary control valve, refer to paragraph 313.

NOTE: System relief valve, pilot relief valve, cylinder cushion (safety) valve and the flow divider relief (safety) valve can be tested after removal by using an injector tester, or a hydraulic hand pump, and the proper adapters. However, bear in mind that the pressures obtained will be toward the low side of the specified pressure ranges due to the low volume of fluid being pumped. The International Harvester test equipment components are as follows: FES64-7-1 (test block), FES64-7-2 (plug), FES64-7-4 (petcock), FES64-7-5 (gage) and FES64-7-6 (hand pump).

If flow meter equipment is available, proceed as follows: Connect meter inlet to quick disconnect that is convenient and secure the outlet hose in the rear frame filler hole. Be sure to fasten outlet hose securely in the filler hole. Start engine and run until fluid is at operating temperature, then actuate auxiliary control valve to direct fluid through the flow meter and note the free flow gpm. This flow should approximate the combined output of both pumps, less the three gallons per minute taken by the flow divider valve for the power steering system. Now slowly close the restrictor valve of the test set and observe the flow meter. If both relief valves are functioning properly, the pressure gage should show the 1550-1600 psi with no appreciable decrease in the flow rate. However, if when the pressure begins to build to the higher pressures, an abrupt drop of approximately 4½ to 7 gpm occurs, it would indicate that the lift (draft and position control) system relief valve is faulty. Similarily, if a drop of approximately 9 to 14 gpm occurs, it would indicate that the auxiliary circuit relief valve in the transfer block is faulty.

295. The relief valve for the lift (draft and position control) system is located in the right front corner of the lift housing as shown in Fig. IH1257. Valve can be removed after removing seat and hydraulic lift housing cover and the procedure for doing so is obvious. Refer to paragraph 294 for testing procedures.

Early relief valves were factory set and staked and could not be adjusted. Later valves have component parts catalogued.

SYSTEM ADJUSTMENTS

Series 606-2606

296. Whenever hydraulic lift unit has been serviced and adjustment is required, the system should be cycled, by using the position control lever, at least ten or twelve times to insure purging any air which might be present in system. All checking and adjusting of the system should be done with a load on the hitch and the hydraulic fluid at operating temperature.

IMPORTANT: Adjustments are made with the top cover of unit removed. The system pressure relief valve located in the right front corner of the unit housing discharges upward when it relieves, therefore, it is necessary to install a shield over the relief valve when operating the system with the top cover removed.

While some early model tractors may not be so equipped, the later model tractors have a drain plug located in the left front corner of unit housing to permit lowering of the fluid level so adjustments can be made without working below the surface of fluid.

Fig. IH1257 — View of hydraulic draft and position control unit with top cover removed.

C. Cylinder
N. Adjusting collar
R. Relief valve
S. Safety (cushion) valve
T. Timing switch
V. Control valve

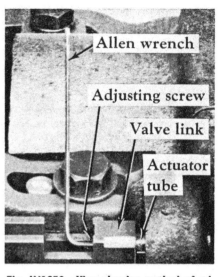

Fig. IH1259—View showing method of adjusting drop poppet actuating rod. Refer to text.

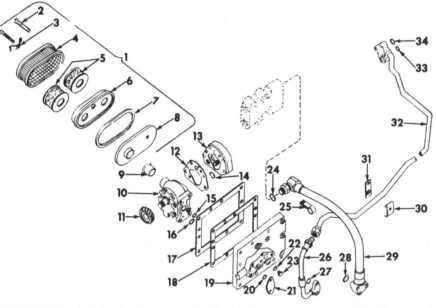

Fig. IH1258—View showing arrangement of hydraulic pumps, pump mounting flange and filter assembly.

1. Filter assembly	8. Support	13. Pump (4½ or 7 gpm)	19. Flange
3. Retainer spring	9. Suction tube	14. "O" ring	20. "O" ring
4. Screen	10. Pump (12 or 17 gpm)	15. "O" ring	21. Cover (no draft control)
5. Elements	11. Drive gear	16. "O" ring	22. "O" ring
6. Retainer	12. Gasket	17. Gasket	23. Plug
7. Gasket		18. Spacer	

Fig. IH1260—Drop poppet actuating rod adjusting screw is located as shown. Valve link has been removed for illustrative purposes.

297. DROP POPPET ADJUSTMENT. Place the draft control lever in the full forward position, raise the hitch to its maximum position with the position control lever, then lower hitch to mid-position. System should go off high pressure with no cycling (hiccups). If unit cycles (hiccups), adjust the drop poppet actuating rod adjusting screw (located between legs of main control valve link) as follows: Again fully raise hitch and lower to mid-position. Turn adjusting screw in until unit begins to cycle (hiccup), then back-out screw until the point

is reached where the unit stops hiccupping. See Figs. IH1259 and IH1260. Now turn the screw out an additional ¾-turn to obtain the proper clearance (0.020) between push rod and check ball.

NOTE: Move the position control handle a small amount after each turn of the adjusting screw to make sure control valve spool is in the normal centered position.

298. POSITION CONTROL LINKAGE ADJUSTMENT. Place draft control lever in full forward position.

Loosen cap screw (B—Fig. IH1261) and slide hydraulic shut-off switch assembly (C) rearward temporarily. Place the position control lever at the offset on forward end of quadrant at which time the hitch should be fully lowered. Check hitch by by-passing the quadrant offset with the position control handle and moving it to the bottom of the slot after loosening lock nut (L) and backing out drop control screw (M). If hitch drops any further, the control valve actuator tube must be turned into the control valve link. See Fig. IH1259. Be sure to tighten lock nut after each adjustment. This adjustment can be checked by moving the position control lever to the rearward position at which

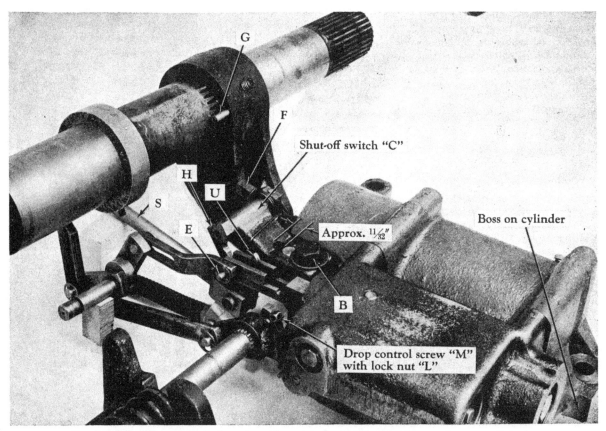

Fig. IH1261—View of the draft and position control unit internal linkage and component parts.

time the rockshaft lift arms should be 45-46 degrees above vertical position. Angle can be determined by measuring from center of lower link pivot pin to the center of the lift link pivot pin. This distance should be $33\frac{3}{16}$ inches with a plus or minus tolerance of ¼-inch.

Note: On some early model tractors, it is permissible to set the position control lever not more than ⅝-inch beyond offset in quadrant to put hitch in the fully lowered position. On these early models, use shims between valve link and actuator tube when adjusting, or install a new actuator tube which has the adjusting nut.

Without moving the position control lever after obtaining the 45-46 degree rockshaft arm position, move the hydraulic switch assembly (C—Fig. IH1261) forward until lever (F) is in contact with pin (G) and lever (H) is in contact with pin (E) and tighten cap screw (B). Distance between end of slot and dowel should be approximately $\frac{11}{32}$-inch as shown in Fig. IH1261. System should go off high pressure when the rockshaft lift arms reach the 45-46 degree above vertical position when hitch is raised with draft control lever.

With position control lever position established as outlined, set bumper stop (Fig. IH1262) on the quadrant so lever contacts it 0.020-0.030 before lever reaches end of its travel and tighten stop securely.

Cycle system several times and recheck the 45-46 degree rockshaft arm position. Rockshaft arms should have approximately 1-inch additional free travel, when lifted manually, after hitch has reached its highest point. If hitch doesn't have the additional 1-inch free travel, recheck the rockshaft lift arm angle and the actuator tube adjustment.

299. DROP CONTROL ADJUSTMENT. Loosen lock nut (L—Fig. IH-1261) and back-out adjusting screw (M). Move position control lever to offset of quadrant, then turn adjusting screw until it just contacts the cam which operates the drop control valve spool. Tighten lock nut.

This adjustment can be checked by moving the draft control lever completely forward. The drop control valve spool should bottom when the lever is about ⅛-inch from full forward.

300. DRAFT CONTROL LINKAGE ADJUSTMENT. Remove cap screw which retains torsion bar flange to bracket and pull torsion bar out until flange clears dowel and bellcrank is free to rotate. Set the position control lever at approximately the offset of the quadrant. Set the draft control lever at the full forward position. Rotate top of bellcrank to the rear until it contacts its stop and hold in this position with a small bar. Rockshaft should not move; however, if it does move, remove pin from draft control rod adjusting nut and turn nut out (counter-clockwise) until bellcrank, with pin installed, will bottom without causing rockshaft to move.

Continue to hold bellcrank against stop and move draft control lever to the rear edge of word "OFF" on the quadrant. The rockshaft arms should raise to the 45-46 degree angular position and the system should go off high pressure. If system does not react as stated, turn draft control rod nut in (clockwise) until operation is correct. Reinstall pin, spring washers and retaining pin. Reinstall torsion bar.

301. CONTROL LEVERS ADJUSTMENT. Position control lever is adjusted by the two locking rings (collars) at inside of control levers and draft control lever is adjusted by the two nuts on the outside of the levers. Procedure for adjusting is obvious. Levers should be adjusted until 4 to 6 lbs. for used friction discs; or 6 to 8 lbs. for new friction discs, is required to move levers. Measurement is taken at control lever knob.

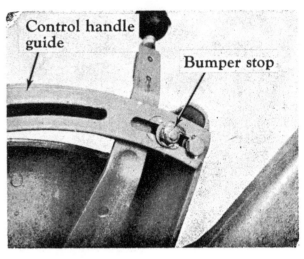

Fig. IH1262—View showing bumper stop. Refer to text for adjustment.

PUMP UNIT

Series 606-2606

The draft and position control hydraulic system pump may be either Cessna or Thompson. Pump capacity may be either 4½ or 7 gpm, however, construction of the pumps remains the same except that pumping gears of the 7 gpm pump are wider. Refer to Figs. IH1263 and IH1264 for exploded views of the two pumps.

302. REMOVE AND REINSTALL. To remove the hydraulic pump, first drain housing, then disconnect the hydraulic lift pressure line and the power steering and auxiliary system pressure line from the pump mounting flange. See Fig. IH1265 for the location of these lines. Mounting flange can now be unbolted and the flange and pump, or pumps, removed. See Fig. IH1266.

On models equipped with both draft and position control and auxiliary system, the draft and position control pump (small) can be separated from the auxiliary system pump (large) after removing the four attaching cap screws. Take care not to damage sealing surfaces when separating pumps.

On all models the auxiliary system (large) pump is attached to the mounting flange with four cap screws. Renew "O" rings each time pump is removed from mounting flange. Cap screws which secure draft and position control pump to auxiliary system pump should be tightened to a torque of 25 ft.-lbs.

NOTE: Beginning with International 606 tractor serial number 1016, a 17 tooth pump drive gear is being used instead of the 15 tooth gear used prior to this serial number. If a new pump (or drive gear which has the 17 teeth) is being installed, it will be necessary to add a spacer to the pump mounting flange and to use a new longer pump suction tube to compensate for the increased diameter gear. It is recommended by the International Harvester Company that when noisy hydraulic pumps are encountered, the 15 tooth pump drive gear be replaced with the 17 tooth gear along with the spacer and new longer

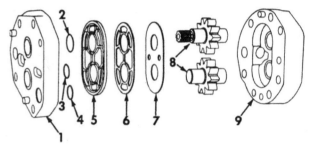

1. Pump cover
2. "O" ring
3. "O" ring
4. "O" ring
5. Diaphragm seal
6. Back-up gasket
7. Diaphragm
8. Pump gears and shafts
9. Pump body

Fig. IH1263 — Exploded view of Cessna draft and position control system pump. Pump may be either 4½ or 7 gpm capacity.

pump suction tube. New pump mounting flanges are also available to accommodate the 17 tooth gear.

When reinstalling the pump (or pumps) and mounting flange on the tractor, make certain that the longer smooth section of the suction tube (24—Fig. IH1266) is pressed completely into the pump. The lip of the seal, which is integral with the tube, should face toward left (pump) side of tractor. Vary the number and thicknesses of the gaskets which are located between the pump mounting flange and the clutch housing to provide a backlash of 0.002-0.022 between pump drive gear and PTO gear. Gaskets are available in thicknesses of 0.011-0.019 and 0.016-0.024.

303. OVERHAUL (CESSNA). With unit removed as outlined in paragraph 302, remove retaining cap screws and separate cover (1—Fig. IH1263) and body (9). Remove "O" rings (2, 3 and 4), diaphragm seal (5), back-up gaskets (6), diaphragm (7) and the gears and shafts (8).

Bushings in body and cover are not available separately. Pump gears and shafts are available only in pairs and when ordering the gears and shafts package, also order an "O" ring and gasket package.

Refer to the following specifications.

4½ GPM Pump
Thickness (width) of gears
(min.) 0.309
O. D. of shafts at bushings (min) 0.685
I. D. of bushings in
body and cover (max.)....... 0.691
I. D. of gear pockets (max.).... 1.719

7 GPM Pump
O. D. of shafts at bushings (min) 0.685
I. D. of bushings in
body and cover (max.)....... 0.691
Thickness (width) of gears
(min.) 0.457
I. D. of gear pockets (max.).... 1.719

When reassembling, use new diaphragm (7), back-up gasket (6) and diaphragm seal (5). Use all new "O" rings. Lubricate gear and shaft assemblies prior to installation. Bronze face of diaphragm (7) must face pump gears and must fit inside the raised rim of the diaphragm seal (5). If necessary, use a blunt tool to work diaphragm seal into position. Tighten body to cover retaining screws evenly and check rotation of pump. Pump will have a slight drag but should turn evenly with no tight spots.

304. OVERHAUL (THOMPSON). With pump removed as outlined in paragraph 302, remove retaining cap

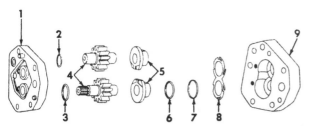

1.	Pump cover
2.	"O" ring
3.	"O" ring
4.	Pump gears and shafts
5.	Bearings
6.	"O" ring
7.	"O" ring retainer
8.	Pressure plate spring
9.	Pump body

Fig. IH1264 — Exploded view of Thompson draft and position control system pump. Pump may be either 4½ or 7 gpm capacity.

Fig. IH1266—View of Cessna pumps after removal of mounting flange. Note how pumps are bolted together. Item (24) is suction tube. Thompson pumps are similar.

Fig. IH1265 — View of pump mounting flange. Small line supplies draft and position control. Large line supplies power steering and auxiliary systems via a flow divider valve.

Fig. IH1267—Seal tube (T) and seal (S) assembly can be removed after torsion bar bracket is off.

screws and separate cover (1—Fig. IH1264) and body (9). Remove "O" rings (2 and 3) from pump cover. Remove gears and shafts (4), bearings (5), "O" rings (6), "O" ring retainers (7) and pressure plate spring (8) from body.

Pump body is not available separately, and if defective, renew complete pump. Pump cover and bushings are not available separately. Pump gears and shafts (4) and bearings (5) are available in pairs only and when ordering pump gears and shafts, bearings and/or pump cover, also order an "O" ring and gasket package.

Refer to the following specifications.

4½ GPM Pump
O. D. of shafts at bushings
(min.) 0.625
I. D. of bearings and
cover bushings (max.) 0.6277
Tickness (width) of gears
(min.) 0.4395
I. D. of gear pockets (max.) ... 1.449

7 GPM Pump
O. D. of shafts at bushings
(min.) 0.625
I. D. of bearings and
cover bushings (max.) 0.6277
Tickness (width) of gears
(min.) 0.6865

I. D. of gear pockets (max.) ... 1.449

When reassembling, use all new "O" rings and gaskets. Lubricate gears and shafts prior to installation and if reusing bearings, reinstall them in their original positions. Tighten body to cover screws evenly and check rotation of pump. Pump will have a slight drag but should turn evenly with no tight spots.

HYDRAULIC LIFT UNIT

Series 606-2606

305. **REMOVE AND REINSTALL.** To remove the hydraulic lift unit from tractor, first remove plug from lower rear right hand corner of lift housing and drain housing. Remove seat, upper lift link and right fender. Remove retainer from right end of draft control bellcrank pin and slide pin to left until it clears draft control nut. Unbolt torsion bar flange from torsion bar bracket and remove torsion bar and bellcrank. Unbolt and remove torsion bar bracket from lift housing.

Note: At this time, bushings in torsion bar bracket, and seal and bearings in draft control rod tube (Fig. IH1267) can be renewed. Control rod

tube can be removed by unscrewing after draft control rod nut is off.

Disconnect hydraulic pump manifold flange at lift housing, and either disconnect manifolds and pipes from auxiliary control valves and transfer block or if desired, remove the valves and transfer block. Disconnect lift links from rockshaft arms. Remove cap screws which retain lift housing to rear center frame, attach chain and hoist and lift unit from tractor.

DRAFT CONTROL CYLINDER AND VALVE ASSEMBLY

Series 606-2606

306. **REMOVE AND REINSTALL.** To remove the cylinder and valve assembly, remove plugs from rear lower right hand corner of lift housing and drain housing. Remove seat, seat brackets and top cover. Remove "C" ring and pin from control valve link and disconnect control linkage from control valve link, move control levers rearward, then unbolt and remove the complete cylinder and valve assembly from housing.

Note: Catch safety switch assembly as mounting cap screws are loosened. Tip forward end of cylinder and valve

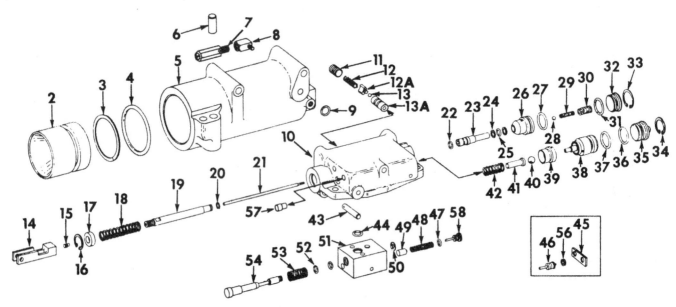

Fig. IH1268—Draft control cylinder and valve assembly showing component parts and their relative positions. Early drop control valves used a retainer and spacer (items 45 and 56) instead of the screw in type shown as item (58).

2. Piston	11. Retaining screw	17. Spring retainer	27. "O" ring	38. Unloading valve	48. Spring
3. Back-up washer	12. Spring	18. Spring	28. Ball	piston	49. Drop valve piston
4. "O" ring	12A. Ball retainer	19. Actuator tube	29. Spring	39. Valve seat	50. Retaining ring
5. Cylinder	13. Check ball	20. Retaining ring	30. Spring	40. Ball	51. Drop valve body
6. Dowel	13A. Flow control	21. Drop valve	31. "O" ring	41. Ball rider	52. "O" ring
7. Relief (safety)	valve	actuating rod	32. Plug	42. Spring	53. Spring
valve	14. Actuator link	22. "O" ring	33. Snap ring	43. Pivot pin	54. Variable orifice
8. Elbow	15. Drop valve	23. Pilot valve seat	34. Snap ring	44. "O" ring	spool
9. "O" ring	adjusting screw	24. Back-up washer	35. Plug	45. Plug retainer	56. Spacer
10. Draft control body	16. Snap ring	25. "O" ring	36. "O" ring	46. Plug	57. Retaining pin
		26. Drop poppet valve	37. "O" ring	47. "O" ring	58. Plug

assembly upward when removing from housing.

NOTE: When reinstalling, it is necessary that the cylinder be seated against the boss in lift housing. Use a small bar to hold cylinder against boss while tightening retaining cap screws. In some cases it may be necessary to reposition the draft control valve.

After unit is installed, refer to paragraph 296 for adjustment. See Figs. IH1268 and IH1269.

307. OVERHAUL. Because of the inter-relation of the cylinder and piston, draft control valve and drop control valve, they will be treated as subassemblies and each will be removed from the complete assembly, serviced and reinstalled.

308. DROP CONTROL VALVE. To remove and overhaul the drop control valve assembly, remove the complete cylinder and valve assembly as outlined in paragraph 306.

Remove the Allen screws which retain drop control valve, lift same from draft control valve and remove "O" ring. Remove "C" retainer from end of variable orifice spool, pull spool from bore and remove the "O" rings from each end of body bore. On early valves, remove end plate (retainer) and remove the spacer, plug and guide (roll pin), spring and piston.

Fig. IH1269 — Cylinder and control valve assembly shown removed. Note that valve link is removed.

C. Cylinder
D. Drop control valve
P. Piston
V. Control valve

Remove "O" ring from plug assembly. See Fig. IH1270. On later valves plate is not used and plug serves as the retainer.

Inspect spool, piston and bores for nicks, burrs, scoring and undue wear. Variable orifice spool and valve should fit bores snugly, yet slide freely. Inspect springs for fractures, distortion or signs of permanent setting. Refer to following specifications and renew parts as necessary.

Use all new "O" rings, reassemble valve by reversing disassembly procedure and install unit on draft control valve.

Variable Orifice Spool Spring
    Free length—In. 1⅞
    Test load lbs. at
        length—In. . . . . 13.5-16.5 @ ⅞

Fig. IH1270—View of drop control valve (early type) with internal parts removed. Late valves use a screw type plug instead of the retainer plate shown.

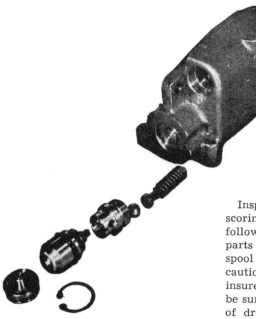

Fig. IH1271 — Unload valve assembly removed from control valve body.

Inspect all parts for nicks, burrs, scoring and undue wear. Refer to the following specifications and renew parts as necessary. Valve body and spool are not available separately. Use caution when renewing "O" rings to insure that proper size is installed and be sure the back-up rings used in bore of drop poppet valve are positioned on each side of the "O" ring. Be sure all check balls and their respective seats are in good condition.

Coat all parts with lubricating oil

Fig. IH1274—Flow control valve and check ball shown removed. Plug has a nylon locking insert.

and reassemble by reversing disassembly procedure. Use new "O" ring and install valve to cylinder assembly. Install drop control valve assembly to draft control valve.

Unloading Valve
Spring free length—In. ....$1\frac{7}{16}$
Test load lbs. at
length—In. ....16.0-17.0 @ $1\frac{1}{16}$

Piston Spring
Free length—In. ....$1\frac{3}{32}$
Test load lbs. at
length—In. .....10.5-12.3 @ 61/64

309. DRAFT CONTROL VALVE. To remove and overhaul the draft control valve, remove the complete cylinder and valve assembly as outlined in paragraph 306. Remove drop control valve from draft control valve, then remove draft control valve from work cylinder and piston assembly.

With valve removed, refer to Fig. IH1271 and remove internal snap ring, then remove plug and unloading valve assembly. Unscrew valve seat and remove ball, ball rider and ball rider spring.

Refer to Fig. IH1272 and remove internal snap ring, then remove plug, drop poppet valve spring (large), pilot valve spring (small), ball and the drop poppet valve and pilot valve seat assembly. Pull pilot valve seat from poppet valve.

Refer to Fig. IH1273 and remove link from control valve, then remove internal snap ring, and remove spring retainer, main valve spring, main valve actuator tube and valve spool. Remove the small retaining ring and pull drop valve actuating rod from actuator tube. Do not remove drop valve adjusting screw (Fig. IH1260) from outer end of actuator tube unless necessary.

Refer to Fig. IH1274 and remove plug, spring, and nylon check valve ball retainer and ball and the flow control valve.

Fig. IH1272—Drop poppet valve assembly removed from control valve body.

Fig. IH1273—Main control valve assembly removed from control valve body. Note drop control valve cam.

Drop Poppet Valve

Spring free length—In. 41/64
Test load lbs. at
   length—In. ....9-11 @ $1\frac{11}{32}$
Check ball spring free
   length—In. 59/64
Check ball spring test load
   lbs. at length—In. ...3.5-4.1 @ ¾

Control Valve

Spring (return) free
   length—In. 2 61/64
Test load lbs. at
   length—In. ......18.4-21.6 @ $1\frac{23}{32}$

Flow Control Valve

Spring free length—In. 1¼
Test load lbs. at
   length—In. ....4.9-5.8 @ $1\frac{1}{16}$

310. WORK CYLINDER AND PISTON. To remove and overhaul the work cylinder and piston, first remove the complete cylinder and valve assembly as outlined in paragraph 306, then remove the draft control valve and drop control valve assembly from work cylinder.

Piston can be removed from cylinder by bumping open end of cylinder against a wood block, or by carefully applying compressed air to the cylinder oil inlet port. With piston removed, the piston "O" ring and back-up washer can be removed.

Inspect piston and cylinder bore for nicks, burrs, scoring or undue wear. Small defects can be corrected by using crocus cloth. Renew parts which are unduly scored or worn. Piston outside diameter is 3.497-3.499 inches. Cylinder inside diameter is 3.500-3.502 inches.

When installing piston "O" ring and back-up washer, be sure "O" ring is toward pressure (closed) end of piston. Coat piston assembly with oil prior to installation in cylinder. Use new "O" ring between cylinder and control valve assembly.

Note: The safety relief (cushion) valve (S—Fig. IH1257) attached to left side of work cylinder can be removed after lift housing top cover has been removed and procedure for doing so is obvious. Relief valve can be bench tested by using an injector tester; however, unit is preset at the factory and is non-adjustable. Renew valve if found to be faulty. Valve is set to relieve at 1650-1900 psi.

311. CONTROL LEVERS AND SHAFTS AND CONTROL LINKAGE. Control levers, quadrant and control lever shafts can be removed as an assembly as follows: Drain hydraulic lift housing and remove seat and lift housing top cover. Remove cylinder and valve assembly as outlined in paragraph 306. Remove control handle knobs, then remove cap screw from one end of control handle guide, loosen other at opposite end and allow guide to hang. Remove quadrant and note position the quadrant was mounted (i.e., which two holes used) so it can be reinstalled the same way. Pull draft control handle rearward, then drive out roll pin at inner end of shaft and remove control lever. Remove snap ring from inner end of position control lever shaft. Unbolt control lever and quadrant support from hydraulic housing and pull support and control levers from housing.

Draft control linkage can now be removed after removing adjusting nut from aft end of draft control rod. Any further disassembly of linkage will be evident after an examination of same. Removal of rockshaft bellcrank and actuating hub will require removal of rockshaft as outlined in paragraph 312.

Oil seal and bearing for control lever shaft can now be removed from lift housing, if necessary.

Any disassembly and/or overhaul required on the control levers and quadrant assembly will be obvious after an examination of the unit. Note that the inner control lever shaft is sealed to the outer control lever shaft by an "O" ring.

NOTE: The eccentric shaft can also be removed at this time. Disconnect internal linkage, if not already done, then remove outer retainer and remove eccentric shaft from inside of housing. Eccentric shaft is fitted with an "O" ring seal. Shaft bushing can also be pulled from lift housing.

When reassembling control levers and drop valve actuating lever to shafts, be sure to align the register marks. Belleville washers on outer end of inner (draft control) shaft are installed as follows: Outer washer, dish toward inside; center washer, dish toward outside; inner washer, dish toward inside.

312. ROCKSHAFT. If rockshaft seal renewal is all that is required, seals can be renewed as follows: Remove control lever quadrant and both lift arms. Use a screwdriver, or similar tool, and pry out old seals. Use a suitable driver and drive new seals in until they bottom. Note that left seal has a smaller inside diameter than the right seal.

To remove the rockshaft, remove the right fender, in addition to the quadrant and rockshaft lift arms. Remove the cylinder and valve assembly as outlined in paragraph 306. Remove Allen screws from actuating hub and bellcrank and slide rockshaft from left to right out of housing, bellcrank and actuating hub. Remove actuating hub key as soon as it is exposed. If actuating hub sticks on rockshaft, either pry against it with a heavy screwdriver, or use a spacer between hub and housing. DO NOT drive on rockshaft without supporting actuating hub as damage to linkage will occur.

Always renew oil seals whenever rockshaft is removed; however, do not install the seals until after the rockshaft is installed. Rockshaft bushings can be removed and reinstalled using a proper sized bushing driver. Outside edge of bushings should be flush with bottom of oil seal counterbore. Inside diameter of left bushing is 2.090-2.095; inside diameter of right bushing is 2.315-2.320. Outside diameter of rockshaft at bearing surfaces is 2.085-2.087 for the left and 2.310-2.312 for the right.

Prior to reassembly, it is recommended that the following identification marks be made even though the rockshaft and rockshaft bellcrank are master splined. These marks will provide visibility and aid in reassembly. Use yellow paint and paint the "V" notch in the actuating hub key, the rockshaft master spline and the allen screw seat in the rockshaft. Also paint a line straight up from the master spline in the rockshaft bellcrank.

Start rockshaft into housing from right side of housing and start actuating hub and rockshaft bellcrank over rockshaft. Align the affixed markings (master splines) of rockshaft and bellcrank and position bellcrank until set screw seat in rockshaft is aligned with set screw hole in bellcrank, then install the Allen screw. Install the actuating hub key and slide the actuating hub over key until the "V" notch is visible through set screw hole, then install the set screw. NOTE: Use a mirror during these operations. Install new oil seals. Install rockshaft lift arms and torque the retaining bolts to 170-190 ft.-lbs.

Complete reassembly by reversing the disassembly procedure.

NOTE: If tractor has been equipped with the longer lift links and the original rockshaft retained, install lift arms two splines up from the present timing mark. If a late type rockshaft is installed, use the upper of the two timing marks when installing lift arms.

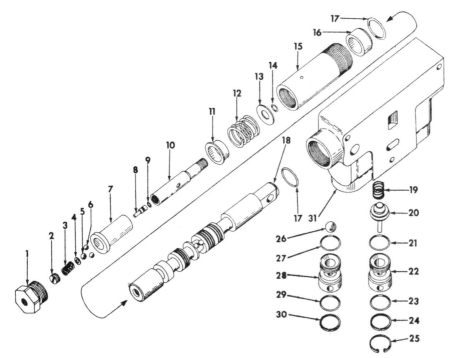

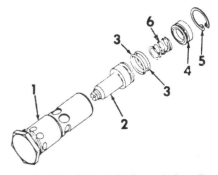

Fig. IH1276—Exploded view of the pilot relief valve which is installed in the transfer block.

1. Plug
2. Relief valve
3. "O" rings
4. Cap
5. Retaining ring
6. Spring

Fig. IH1275—Exploded view of the hydraulic auxiliary control valve used for remote cylinders. Items 26 through 30 are used in early type valves. Items 19 through 25 are used in late type valves.

1. Cap	7. Position control sleeve	15. Sleeve
2. Plug	8. Unlatching piston	16. Retainer
3. Spring	9. "O" ring	17. "O" ring
4. Detent actuating ball washer	10. Actuator	18. Spool
5. Detent actuating ball	11. Spring retainer	19. Poppet spring
6. Detent balls	12. Centering spring	20. Poppet
	13. Washer	21. "O" ring
	14. "O" ring	22. Retainer
		23. "O" ring

24. Back-up washer
25. Snap ring
26. Check ball
27. "O" ring
28. Retainer
29. "O" ring
30. Back-up washer
31. Body

AUXILIARY CONTROL VALVE

Series 606-2606

313. **R & R AND OVERHAUL.** To remove the auxiliary control valve, or valves, first remove right fender, then remove control lever knobs and quadrant assembly.

NOTE: Be sure to identify the cap screw holes in the quadrant before removing the cap screws. Mounting position of quadrant differs between Farmall and International model tractors.

Remove banjo bolts and disconnect manifolds from control valves. Remove through bolts and pull cover and control valves from transfer block.

When reinstalling, torque mounting bolts to 20-25 ft.-lbs. Do not overtighten mounting bolts as valve body may be distorted and valve sticking could result.

To disassemble, use Fig. IH1275 as a guide. Remove control handle and bracket. Remove end cap (1), then unscrew the actuator (10) and remove the actuator and detent assembly. Remove sleeve (15) and pull balance of parts from body. Check ball and re-

tainer can be removed at any time. Note the difference between early check valves (items 26 through 30) and late check valves (items 19 through 25). On late type check valves, snap ring (25) must be removed before retainer (22) can be removed.

NOTE: Some valves do not include the detent assembly. When disassembling these valves, sleeve (15) must be removed before removing actuator (10).

In addition, industrial valves have a circuit relief valve located directly below sleeve (15) and valve can be removed at any time.

Detent (3, 4, 5 and 6) can be disassembled after removing plug (2). Push unlatch piston (8) out of actuator (10) by using a long thin punch.

Inspect all parts for nicks, burrs, scoring and undue wear and renew parts as necessary. Spool (18) and body (31) are not available separately. Check detent spring (3) and centering spring (12) against the following specifications.
Detent spring

Free length—In. ....$1\frac{1}{16}$

Test load lbs. at

length—In. .....23.5-28.5 @ 45/64
Centering spring

Free length—In. ....$2\frac{5}{16}$

Test load lbs. at

length—In. ....26.5-33.5 @ 1 7/64

Use all new "O" rings and reassemble by reversing the disassembly procedure. Detent unlatching pressure is adjusted by plug (2). Unit must unlatch at not less than 1000 nor more than 1250 psi. The circuit relief valve on industrial valves is a cartridge type with the pressure setting stamped on end of body. Faulty relief valves are corrected by renewing the complete unit. Be sure filter in end cap (1) is clean (no paint) and in satisfactory condition.

TRANSFER BLOCK

Series 606-2606

314. **R&R AND OVERHAUL.** To remove the transfer block, first remove the auxiliary control valves as outlined in paragraph 313. Disconnect the supply line, then remove the two cap screws and the socket head screw and pull transfer block from lift housing.

Check ball seat on tractors prior to serial number 2109 can be removed by using two small screw drivers to pry it out of transfer block. Removal of check ball retainer is obvious.

On tractors serial number 2109 and up, a poppet type check valve assembly has replaced the ball type and service procedure is obvious.

Pilot relief valve is removed as a unit by unscrewing plug (1—Fig. IH-1276). Note the relief valve opening pressure which is stamped on valve should it be necessary to bench test the valve. Plug (1) and relief valve cartridge (2) are not available separately and if either is defective, renew complete valve.

NOTES